Progress in Photothermal and Photoacoustic Science and Technology

Progress in Photothermal and Photoacoustic Science and Technology

Andreas Mandelis, Editor

Volume I

Principles and Perspectives of Photothermal and Photoacoustic Phenomena

Volume II

Non-Destructive Evaluation

NON-DESTRUCTIVE EVALUATION (NDE)

Volume II in the Series

Progress in Photothermal and Photoacoustic Science and Technology

Andreas Mandelis
EDITOR

Director, Photoacoustic and Photothermal Sciences Laboratory
Department of Mechanical Engineering
University of Toronto
Toronto, Canada

PTR Prentice Hall
ENGLEWOOD CLIFFS, NEW JERSEY 07632

Acquisitions editor: Mike Hays
Editorial assistant: Kim A. Intindola
Cover design: Bruce Kenselaar
Cover design director: Eloise Starkweather-Muller
Art production manager: Gail Cocker-Bogusz
Production coordinator (buyer): Alexis Heydt

 © 1994 Prentice Hall
Prentice-Hall, Inc.
A Paramount Communications Company
Englewood Cliffs, New Jersey 07632

The publisher offers discounts on this book when ordered in bulk quantities. For more information, contact Corporate Sales Department, PTR Prentice Hall, 113 Sylvan Avenue, Englewood Cliffs, NJ 07632. Phone: 201-592-2863; FAX: 201- 592-2249.

All rights reserved. No part of this book may be reproduced, in any form or by any means, without permission in writing from the publisher.

Printed in the United States of America

10 9 8 7 6 5 4 3 2 1

0-13-147430-8

Prentice-Hall International (UK) Limited, London
Prentice-Hall of Australia Pty. Limited, Sydney
Prentice-Hall Canada Inc., Toronto
Prentice-Hall Hispanoamericana, S.A., Mexico
Prentice-Hall of India Private Limited, New Delhi
Prentice-Hall of Japan, Inc., Tokyo
Simon & Schuster Asia Pte. Ltd., Singapore
Editora Prentice-Hall do Brazil, Ltda., Rio de Janeiro

International Advisory Board

A.C. Boccara, *Paris, France*
S.E. Braslavsky, *Mulheim, Germany*
G. Busse, *Stuttgart, Germany*
D. Cahen, *Rehovot, Israel*
H.J. Coufal, *San José CA, USA*
G.J. Diebold, *Providence RI, USA*
J.M. Harris, *Salt Lake City UT, USA*
P. Hess, *Heidelberg, Germany*
R.E. Imhof, *Glasgow, United Kingdom*
P. Korpiun, *Garching, Germany*
M. Luukkala, *Helsinki, Finland*
J.F. McClelland, *Ames IA, USA*
R.M. Miller, *Merseyside, United Kingdom*
J.-P. Monchalin, *Boucherville Que., Canada*
J.C. Murphy, *Laurel MD, USA*
R.A. Palmer, *Durham NC, USA*
J. Pelzl, *Bochum, Germany*
A. Rosencwaig, *Fremont CA, USA*
B.S.H. Royce, *Princeton NJ, USA*
T. Sawada, *Tokyo, Japan*
M.W. Sigrist, *Zürich, Switzerland*
A.C. Tam, *San José CA, USA*
R.L. Thomas, *Detroit MI, USA*
H. Vargas, *Campinas, Brazil*
S.-Y. Zhang, *Nanjing, China*
V.P. Zharov, *Moscow, USSR*

Aims and Scope of the Series

The Series is devoted to the presentation of concise, authoritative reports of progress by leading scientists in the existing and/or rapidly developing interdisciplinary research areas (themes) of Photothermal and Photoacoustic Science and Technology. Due to the intensely interdisciplinary character of the field since its modern day inception, the goals of the series are:

- To serve as an international forum for enhancing and focusing on the understanding and significance of important progress in diverse areas of applications;

- To keep active workers in, and casual users of, the Photothermal and Photoacoustic field abreast of new progress in areas outside their own specialty;

- To focus the interest of scientists and engineers from other disciplines on the state-of-the-art research and on the potential importance of Photothermal and Photoacoustic Science and Technology, as a multi-disciplinary tool, in their own field(s).

The series covers the entire range of Photothermal and Photoacoustic activities through specially designated volumes, each of which is dedicated to a particular well-defined research theme, spanning a host of physical, chemical, engineering, biological, and (bio)medical disciplines. It constitutes a continuous reference on state-of-the-art research developments, in an attempt to eliminate the existing lack of continuity and focus in several theme areas of common origin and discipline. It is hoped that the series will contribute in a major way towards strengthening the interactions among sub-disciplines, will stimulate further progress, and will catalyze exciting new directions and applications.

All invited contributions are fully peer-reviewed by internationally renowned leaders in the field of Photothermal and Photoacoustic Science and Technology.

Contents

Editorial Preface
List of Contributors
Preface to Volume II

1. HISTORICAL PERSPECTIVE

ALLAN ROSENCWAIG

I. Introduction 2
II. Ultrasonics and Thermoacoustics 3
III. Defect Detection 4
IV. Layers and Films 7
V. PhysicaL PropertieS 7
VI. Novel Techniques 8
VII. Conclusions 10
VIII. References 10

2. REAL-TIME THERMAL WAVE IMAGING

L.D. FAVRO, P.K. KUO AND R.L. THOMAS

I. Introduction 24
II. Box-Car Thermal Wave Imaging 29
III. Pulsed Thermal-Wave Tomography 35
IV. Inverse Scattering and Object Reconstruction 38
V. Lock-In Video Thermal Wave Imaging 40
VI. The Flying Spot Camera 42
VII. References 47

3. THEORY OF THERMAL WAVE SURFACE TEMPERATURES AND SURFACE DISPLACEMENTS FOR BULK AND LAYERED SAMPLES IN ONE AND THREE DIMENSIONS

JON OPSAL AND ALLAN ROSENCWAIG

I. Introduction 54
II. Theory 54
III. Discussion 68
IV. References 69

4. **THERMAL WAVE MONITORING AND IMAGING OF ELECTRONIC MATERIALS AND DEVICES**

   ALLAN ROSENCWAIG

   I. Introduction 74
   II. Experimental Methodology 75
   III. Ion Implant Monitoring 77
   IV. Etch Monitoring 85
   V. Metallization Monitoring 86
   VI. Polycrystalline and Amorphous Silicon 88
   VII. Imaging of Subsurface Silicon Defects 90
   VIII. Imaging of Subsurface Defects in Metal Lines 99
   IX. Conclusions 106
   X. References 106

5. **PHOTOTHERMAL RADIOMETRY OF SEMICONDUCTORS**

   STEPHEN SHEARD AND MIKE SOMEKH

   I. Introduction 112
   II. PTR Signal Generation 116
   III. The PTR Microscope 126
   IV. Semiconductor Assessment 129
   V. Summary 147
   VI. References 149

6. **PHOTOACOUSTIC AND PHOTOTHERMAL CHARACTERIZATIONS OF SEMICONDUCTOR SUPERLATTICES AND HETEROJUNCTIONS**

   SHU-YI ZHANG AND TSUGUO SAWADA

   I. Introduction 152
   II. Fundamental Features of Semiconductor Heterostructures 153
   III. Photoacoustic and Photothermal Spectroscopic Characterization of Heterostructures 157
   IV. Photo-Modulated Reflectance Characterization of Heterostructures 164
   V. Thermal Properties of Heterostructures 172
   VI. Non-Destructive Testing of Heterostructures by Picosecond-Laser-Ultrasonics 177

VII. Conclusions 180
VIII. References 181

7. PHOTOTHERMAL RADIOMETRY FOR NDE

R. E. IMHOF, B. ZHANG AND D. J. S. BIRCH

I. Introduction 186
II. Theoretical Background 190
III. Applications 202
IV. Conclusions 227
V. References 227

8. SPATIALLY RESOLVED DETECTION OF MICROWAVE ABSORPTION IN FERRIMAGNETIC MATERIALS

JOSEF PELZL AND OTFRIED VON GEISAU

I. Introduction 238
II. Experimental Aspects 240
III. Applications to Ferrimagnetic Materials 261
IV. Summary 311

9. X-RAY THERMAL-WAVE NON-DESTRUCTIVE EVALUATION

TSUTOMU MASUJIMA AND EDWARD M. EYRING

I. Introduction 322
II. X-Ray Absorption 322
III. X-Ray Photoacoustic Signals and Instrumentation 325
IV. The "Semi-Pulse" X-Ray Photoacoustic Method 328
V. X-Ray Photoacoustic Imaging 331
VI. Conclusions 337
VII. References 337

EDITORIAL PREFACE

The field of photothermal science and technology has reached a level of maturity and diversity such that it is quite difficult for an individual researcher to be aware of many of the new and important developments, even in his (or her) general domain of applications. Fortunately, there exist specific publications, mostly journals, which carry the majority of research articles in a given sub-field, spanning the spectrum of activities from Physics to Plant Science to Medicine. Keeping up-to-date with the progress in this Field is, however, a difficult task if it must be done through random journal publications, precisely due to its extraordinary diversity. The necessity for texts in monograph or reference from, which present photothermal information in a (more or less) coherent manner was felt in the early (largely photoacoustic) community almost immediately after the modern day rebirth of the Field in the seventies, with the first reference text appearing in 1977 (Y-H. Pao, Ed., "*Optoacoustic Spectroscopy and Detection*", Academic, New York). That book, besides being of great importance in consolidating the definition of the Field as an intellectual discipline in its own right, is, in addition, of considerable historical value, as reflective of the wondrous synergy among vastly heterogeneous components. These constituents from many, nominally unrelated, yet crucially important disciplines came together to define Photoacoustics, largely due to the pioneering inspirations of the early expert, such as P.C. Claspy, L.B. Kreuzer, C. Forbes Dewey, Jr., A. Rosencwaig, and M.B. Robin, among others.

As the Field evolved, two books contributed to the definition of the continuum now known as Photoacoustic and Photothermal Science and Technology (A. Rosencwaig, "*Photoacoustics and Photoacoustic Spectroscopy*", Wiley, New York, 1980; and V.P. Zharov and V.S. Letokhov, "*Laser Optoacoustic Spectroscopy*", Springer-Verlag, Berlin, 1986). Meanwhile, during the past decade the generation of young workers who grew into the Field through its formative years (among whom I count myself) had the enormous opportunity to help develop the roadways towards its (sometimes controversial, often ill-understood by outsiders) present day status as an interdisciplinary set of richly interwoven diagnostic principles, tools and techniques, capable of successfully dealing with the stringent requirements of a rapidly evolving high-technology world. The long-awaited specificity of Photothermal Science in addressing "real-life" problems, an unequivocal sign of its "coming of age", was amply demonstrated by the emergence of specialized texts in the late eighties (A. Mandelis, Ed., "*Photoacoustic and Thermal Wave Phenomena in Semiconductors*", North-Holland, New York, 1987; J.A. Sell, Ed., "*Photothermal Investigations of Solids and Fluids*", Academic, San Diego, 1989; P. Hess, Ed., "*Photoacoustic, Photothermal and Photochemical Processes in Gases*", Springer-Verlag, Heidelberg, 1989; and

P. Hess, Ed., *"Photoacoustic, Photothermal and Photochemical Processes at Surfaces and Thin Films"*, Springer-Verlag, Heidelberg, 1989). At the time of their appearance, these texts were important in outlining the disciplinary boundaries of several sub-fields of Photoacoustic and Photothermal Science and in describing the most valiant efforts of the Community to develop its nascent technologies, the next unequivocal sign of the coming of age. More recent book publications in the form of Proceedings in the Field include "Photoacoustic and Photothermal Phenomena", Springer-Verlag, Heidelberg, 1989; P. Hess and J. Pelzl, Eds., "Photoacoustic and Photothermal Phenomena II", Springer-Verlag, Berlin, 1990; J.C. Murphy, J.W. Maclachlan Spicer, L.C. Aamodt and B.S.H. Royce, Eds.; and finally "Photoacoustic and Photothermal Phenomena III", Springer-Verlag, Berlin, 1992; D. Bicanic, Ed.

In the diversified nineties, faced with the prolific branching of the Field, it is necessary for the photothermal community to be able to keep abreast of the further developments within the established and emerging sub-fields by focusing on significant new knowledge and applications, impacting the science and technology of the given sub-discipline. A recent cursory search in the extent of the photothermal activity spectrum revealed at least thirty areas of penetration/application: analytical chemistry, biology and bio-medicine, electronic materials, ultrasonic photoacoustics, phase transitions, optical materials, non-destructive evaluation, microscopy and imaging, spectroscopy of gaseous and condensed phases, thermal and mass transport, instrumentation, trace gas and pollutant sensing, agriculture, thin layer and film diagnostics, photothermal-based devices, photochemistry, photobiology and environmental science, ultrafast dynamics, (photo) electrochemistry, non-linear phenomena, thermal properties, adsorption and surface science and non-conventional methodologies (this list os not meant to be complete). It is just as important to communicate to workers outside the Field its rich and rewarding potential in terms readily comprehensible within their own scientific culture. It is my contention that further breakthroughs can and will be made when attention is paid to potential areas of application hitherto nominally irrelevant to photothermal treatment.

It is clear that the materialization of this goal cannot be engineered by randomly appearing textual review information, but that a continuing and concerted effort by leaders of the Community is needed to promote further breakthroughs and to cross-fertilize the knowledge base of the photoacoustic and photothermal discipline with the roots of other disciplines. For this purpose, our Series, largely evolving through the collective wisdom of our International Advisory Board (of three-year rotating tenure) purports to provide the vehicle in which the people who have actively "made it happen" will ensure that critical information and progress in their area of expertise is effectively, efficiently and concisely communicated to the interested reader.

Perhaps its most significant legacy will be when any of the distinguished authors, through critical discussion, succeeds in planting the seed of a farfetched idea to another scientist who can make it come true.

Andreas Mandelis
Toronto, April 1993

List of Contributors to Volume II

DAVID J.S. BIRCH
　　Department of Physics and Applied Physics, Strathclyde University, Glasgow G4 0NG, Scotland

EDWARD M. EYRING
　　Department of Chemistry, University of Utah, Salt Lake City, UT 84112, U.S.A.

LAWRENCE D. FAVRO
　　Institute for Manufacturing Research and Department of Physics, Wayne State University, Detroit, MI 48202

ROBERT E. IMHOF
　　Department of Physics and Applied Physics, Strathclyde University, Glasgow G4 0NG, Scotland

PAO KUANG KUO
　　Institute for Manufacturing Research and Department of Physics, Wayne State University, Detroit, MI 48202

TSUTOMU MASUJIMA
　　Institute of Pharmaceutical Sciences, Hiroshima University, School of Medicine, Kasumi 1-2-3, Hiroshima 734, Japan

JON OPSAL
　　Therma-Wave, Inc., Fremont, CA 94539

JOSEF PELZL
　　Institut für Experimentalphysik, AG Festkörperspektroskopie, Ruhr-Universität Bochum, D-44780 Bochum, FRG

ALLAN ROSENCWAIG
　　Therma-Wave, Inc., Fremont, CA 94539

TSUGUO SAWADA
　　Department of Industrial Chemistry, Faculty of Engineering, The University of Tokyo, Tokyo, Japan

STEPHEN SHEARD
　　University of Oxford, Department of Engineering Science, Parks Road, Oxford OX1 3PJ, U.K.

MIKE SOMEKH
 University of Nottingham, Department of Electrical and Electronical Engineering, University Park, Nottingham NG7 2RD, U.K.

ROBERT L. THOMAS
 Institute for Manufacturing Research and Department of Physics, Wayne State University, Detroit, MI 48202

OTFRIED VON GEISAU
 Institut für Experimentalphysik, AG Festkörperspektroskopie, Ruhr-Universität Bochum, D-44780 Bochum, FRG

BU-FA ZHANG
 Department of Physics and Applied Physics, Strathclyde University, Glasgow G4 0NG, Scotland

SHU-YI ZHANG
 Laboratory of Photoacoustic Science, Institute of Acoustics, Nanjing University, Nanjing, The People's Republic of China

PREFACE TO VOLUME II

Non-Destructive Evaluation is the subject of this, the second volume of the Series. Now a mature sub-field of Photothermal and Photoacoustic Studies, NDE has engaged the interest of many investigators in recent years. The aspects of NDE presented in this volume were chosen with a view to summarize and critically discuss information not heretofore coherently treated in the scientific and technical literature.

Each volume in this series opens with a historical overview of the subject authored by well-known and respected authorities who played an important role in shaping the historical development of the broader "picture" behind the sub-field in question. Allan Rosencwaig presents his perspective on the historical background and the major strengths of Thermal Wave and Photoacoustic NDE. Subsequent chapters discuss applications to NDE which are considered central in defining the domain and the range of this important sub-field. An obvious absence is that of Ultrasonic Laser Photoacoustics. Readers of this volume should note that this area has been fully covered by C.B. Scruby and L.E. Drain's excellent book titled "Laser Ultrasonics", Adam Hilger, Bristol, 1990.

Furthermore, certain essential aspects of NDE dealing with thermal wave imaging, characterization and microscopy were dealt with in detail in volume I by G. Busse, A. Mandelis, M. Munidasa and H.G. Walther. These particular topics have not been repeated in volume II. The presentation level of the Series is of research quality. It is intended to be used as reference material, as well as for supplementing upper level graduate courses on various subjects of photothermal and photoacoustic science and technology.

At the time of the publication of volume II two major changes are occurring in this Series: First, is the expiration of the three-year tenure of the first International Advisory Board (IAB). I wish to thank all twenty-six colleagues for their seminal contributions to the establishment and direction of the Series and to commend the high quality of the peer reviews marked by the acceptance and publication of all our chapters in volumes I and II. I will be looking forward to the selection of the new IAB by the members of the out-going IAB and to developing a working relationship with them for the next three years. Second, the recent rationalization of the North American operations by Elsevier Science Publishing Co., Inc., has dictated the transfer of their book series to other North American publishers. Prentice-Hall has kindly undertaken the continuation of the publication of our Series.

On behalf of the IAB, I wish to thank all the contributors to this volume for their enthusiasm and prompt response to their refereeing duties.

The contributions of several referees outside the IAB, necessitated by the particular specialty of some chapters, have been very valuable and are gratefully acknowledged. Special thanks are due to the word-processing and administrative staff of the Department of Mechanical Engineering, University of Toronto, Mesdames Guida Néné, Wendy Smith and Margaret Tompsett, for their patience, perseverance and expert processing of the manuscript. I further wish to acknowledge the support of Professor D. McCammond, Chairman of the Department of Mechanical Engineering of the University of Toronto, for kindly putting at our disposal the Department's word-processing, laser-printing and print-shop facilities for the processing and printing of the entire book.

Finally, I wish to acknowledge the understanding of my wife Nancy and my daughters Alexandra and Nicole throughout the lengthy editorial process of this volume.

Andreas Mandelis
Toronto, April 1993

Volume II

**NON-DESTRUCTIVE
EVALUATION (NDE)**

HISTORICAL PERSPECTIVE

Allan Rosencwaig

Therma-Wave, Inc.
Fremont, CA 94539

I.	INTRODUCTION	2
II.	ULTRASONICS AND THERMOACOUSTICS	3
III.	DEFECT DETECTION	4
	1. Gas Microphone and Piezoelectric Methods	4
	2. Optical Beam Deflection	5
	3. Photothermal Radiometry	5
	4. Modulated Reflectance	6
IV.	LAYERS AND FILMS	7
V.	PHYSICAL PROPERTIES	7
VI.	NOVEL TECHNIQUES	8
	1. Particle Beams	9
	2. X-Rays	9
	3. Microwaves	9
	4. Radio Frequency	10
VII.	CONCLUSIONS	10
VIII.	REFERENCES	10

I. INTRODUCTION

Photothermal and photoacoustic science and technology, or more broadly the subject of thermal waves and related phenomena, now includes many diverse areas - physics, materials science, chemistry, biology, medicine and non-destructive evaluation. The area of non-destructive evaluation, or NDE, is the subject of this volume, and is one of the fastest growing areas in thermal wave research.

To many researchers not involved in this field the term non-destructive evaluation connotes routine ultrasonic or radiological examination of machine parts in a failure analysis laboratory. This is unfortunate since the field has changed dramatically in the last 20-30 years. To appreciate the wide scope of modern-day NDE, we might first consider the general applications of scientific and technological methodologies and instrumentation. One major application is in the field of basic research, investigating new phenomena and materials and making more precise measurements of fundamental physical parameters. Another major area is in analysis, particularly chemical analysis. A third major area is in clinical diagnostics. Although there is thermal wave research activity in all three of these areas, these are not considered within the scope of the NDE field.

Non-destructive evaluation can be defined as the measurement, inspection, or analysis of materials and processes as part of a manufacturing of fabrication cycle. Modern manufacturing processes rely heavily on many measurements, inspections, and analyses. Some of these are performed in an R&D setting devoted to process development. Others are performed in the more traditional failure analysis and reliability laboratory testing a manufactured product after the manufacturing cycle is completed or after field use. However, the fastest growing area for NDE measurements is during the manufacturing cycle itself. Thus there is considerable interest in off-line process monitoring, in in-line monitoring after a critical process step, and even in *in-situ* monitoring during the actual process.

As the complexity and control requirements of manufacturing processes increase, the importance of metrology, inspection and analysis in all phases of the manufacturing cycle also increases. Thus modern day NDE is no longer limited to the failure analysis laboratory. Instead it also includes process development and off-line, in-line and *in-situ* monitoring as well. The resulting requirements on metrology, analysis and inspection technologies are becoming so demanding that the NDE field is now a significant source of new concepts, theories and technologies in these areas.

Thermal wave techniques have proven to be very useful in NDE [1] and there has been considerable and growing activity in this area. It is not possible, in this historical perspective, to cover adequately such a large and growing body of research. I have, most probably, inadvertently neglected several important contributions, and for this I offer my sincere apologies.

Nevertheless, I hope that with this historical perspective I am able to convey the scope and excitement of thermal wave concepts, experiments and applications in non-destructive evaluation.

II. ULTRASONICS AND THERMOACOUSTICS

Thermal wave physics plays an important role in acoustic and ultrasonic technologies when the acoustic or ultrasonic waves are generated through the interaction of a pulsed or modulated energy beam (eg. a laser) with the sample.

Ultrasonics has long been a staple of NDE methodologies, originally for failure analysis applications, but more recently for process control and off-line monitoring as well. Ultrasonics is useful because of its ability to penetrate opaque objects, determine microstructural parameters such as grain size, and process variables such as internal temperature distributions, and to detect internal structural discontinuities such as cracks, voids and inclusions.

Ultrasonic waves are commonly generated and detected with piezoelectric transducers in contact with the sample through a suitable coupling agent such as water or gel. Although piezoelectric generation is quite efficient and piezoelectric detection is very sensitive, the need for contact to the sample is a key limitation. Piezoelectric transducers are fragile, require coupling agents and cannot tolerate high temperatures. Thus it is difficult to use them in severe environments, with samples having complex shapes or at elevated temperatures. In addition, piezoelectric transducers have limited bandwidth and often suffer from distorted frequency response due to acoustic ringing.

The primary advantage of laser-generated ultrasound is its non-contact nature. In fact, when combined with laser detection of ultrasound, the entire process becomes non-contact in nature. In addition, the laser based technique can be used at high temperatures. Laser ultrasound can thus be used as an in-process monitor, can easily be adapted to samples with complex shape and is able to provide high frequency bandwidth (i.e. ultrashort) pulses free from ringing artifacts. The major limitation for laser generation of ultrasound is its relatively low efficiency and for laser detection the fact that this method is several orders of magnitude less sensitive than piezoelectric detection.

The use of pulsed lasers to generate ultrasonic waves was first proposed by White [2] and Askar'yan *et al* [3] in 1963. Initial experiments to demonstrate the potential of this technique to NDE were performed by von Gutfeld and Melcher in 1977 [4]. Since then the topic of laser generated ultrasound has been studied both theoretically and experimentally by many authors, and has been reviewed by Hutchins [5] and Tam [6]. Although there are several possible mechanisms involved in the laser generation of ultrasound, including radiation pressure, electrostriction, changes of state (ablation, plasma formation, cavitation, etc.) the dominant mechanism under usual experimental conditions is thermoelastic expansion [7-9], and hence the importance of thermal waves. Laser-generated ultrasound has found several applications in the NDE field. These include: detection of subsurface defects in metals [10]; evaluation of bonding integrity [11,12]; inspection of composite materials [13,14]; measurement of coating and layer thickness [15,16]; determination of grain size in metals and ceramic materials [17,18]; determination of internal temperatures through the measurement of acoustic velocities in materials under high temperature processing [19,20]; determination of elastic constants [21]; and the monitoring of welding [22], ceramic sintering [23], and metal fabrication [24] processes. We can anticipate that the use of laser ultrasound will continue to grow as the technology is improved, and in particular as means are found to increase the efficiency of laser ultrasound generation and the sensitivity of laser detection. Some of the techniques being developed to achieve these improvements include frequency bandwidth narrowing [25], acoustic focusing and beam steering using phase array techniques [26,27] and a number of more sensitive detection methods [28].

Laser or optically generated thermoacoustics differs from laser-generated ultrasound in that the acoustic waves produced are generally cw and at much lower frequencies. At these frequencies the entire sample undergoes various types of vibrations depending on the sample geometry, support configuration, frequency etc. [29]. This technique has been used for testing the delamination of layered materials [30], imaging of vibrational modes [31], inspection of solder joints [32], evaluation of bonding integrity [12], and for the evaluating of various aspects of small mechanical structures [33].

III. DEFECT DETECTION

A major application of thermal wave techniques in NDE has been for the detection and imaging of defects, particularly subsurface defects. Almost all of the various thermal wave methods have been used for this application. Below are some examples of how these various thermal wave methods have been employed for the detection and imaging of subsurface defects.

(1) Gas Microphone and Piezoelectric Methods

The gas microphone method for performing thermal wave analysis on solids was first developed by Rosencwaig [34] as an extension of the extensive prior work with gas microphone cells on the optoacoustic effect in gases [35,36]. As the oldest thermal wave methodology, it is not surprising that the gas microphone technique was the first to be applied to NDE applications such as film thickness measurements [37]. Subsurface defect imaging using the gas microphone method was first demonstrated by Wong et al in 1978 [38] and refined further by a number of investigators [39-41].

Rosencwaig and Busse [42] demonstrated that the phase images were often better for detecting subsurface features since they were essentially insensitive to the surface optical properties. More recent work has been reviewed by Zhang and Chen [43]. See also Chapter 6 in this volume by Zhang and Sawada. The gas microphone method however suffers from two serious limitations. First, the inconvenience of having to mount the sample in a sealed cell. Secondly, the inability of this technique to detect an important class of defects, closed vertical cracks, which can only be detected with those thermal wave methods that are sensitive to lateral heat flow [44].

Piezoelectric detection of the thermoacoustic signal has also been employed initially by von Gutfeld and Melcher [4] in 1977 and since then by many other authors. Applications include layer thickness measurements [45], lattice damage [46] and defects in semiconductor materials [47,48]. Although the piezoelectric method is quite sensitive for the detection of the thermoacoustic signal, the need for a coupling agent between the sample and transducer has limited the applicability of this method, except for the case of particle beam excitation as discussed later.

(2)   Optical Beam Deflection

A thermal wave detection and imaging technique that is non-contact and furthermore is sensitive to lateral heat flow and thus can detect closed vertical cracks utilizes the optical beam deflection or mirage method, first developed by Boccara, Fournier and Badoz [49] and by Murphy and Aamodt [50,51] in 1980. This technique has been used extensively for defect detection because of its simplicity and its contactless nature. Its ability to detect closed vertical cracks has been demonstrated by Lin *et al* [52] and others.

(3)   Photothermal Radiometry

Photothermal radiometry is a method that was initially developed in the NDE community prior to the current interest in thermal waves. This NDE field, commonly known as thermography, has been reviewed by Vavilov and Taylor [53] and by Reynolds [54]. Thermal wave researchers have emphasized and further developed the transient and modulated versions of photothermal radiometry. Because of its simplicity, its contactless nature and the availability of IR video cameras, this technique has become the most popular thermal wave method for detecting and imaging subsurface defects that are not microscopic in size; see Chapter 7 in this volume by Imhof.

Some of the initial thermal wave experiments utilizing IR detectors were performed by Nordal and Kanstad [55] and by Busse [56] in 1980. The early experiments on subsurface defects [57-66] employed single infrared detectors and obtained spatial information by suitable scanning of either the heating source or the detector. More recent photothermal experiments often involve the use of infrared video cameras and scanners so as to obtain two-dimensional photothermal images of the samples. The combination of synchronous phase lock-in [67,68] and box-car integration methods [69] with these infrared cameras allows for the rapid inspection of large objects and provides valuable information about the depth of defects beneath the sample surface [70]. See also Chapter 2 by Favro, *et al.* in this volume.

Considerable information about subsurface defects can also be obtained from simpler photothermal systems that operate by studying the time evolution of a photothermal signal using pulsed or step heating functions [71-73]. In fact, some photothermal systems are simple enough that they can be used in the field and in hostile industrial environments [74,75].

(4) Modulated Reflectance

Thermal wave techniques have also been particularly useful where measurements need to be performed at high spatial resolution, of the order of 1 micron, or on very thin films. This is particularly true for NDE applications in the semiconductor industry where measurements must be made on very thin layers and on a microscopic scale in a totally non-contact manner.

The most practical methods to perform such measurements involve the use of highly focused laser beams that both generate and detect thermal waves at the front surface of a semiconductor wafer. High resolution thermal wave experiments have been performed with both modulated surface displacement [76,77] and modulated surface deflection or distortion [78,79] techniques. However the most widely used method for such high resolution applications in the semiconductor area is based on the modulated reflectance technique developed by Rosencwaig *et al* [80,81] in 1985. See also Chapter 3 by Opsal and Rosencwaig and Chapter 4 by Rosencwaig in this volume.

Using this technique it has been possible to perform routine in-process monitoring of the ion implantation process [82-84] and of the plasma etch process [85,86]. Both of these applications involve the detection of process induced damage or defects in the crystalline silicon substrate. Other subsurface defect applications include the imaging of individual dislocations, stacking faults, and precipitates in the silicon substrate [87-90] and the detection and imaging of microscopic voids, notches and precipitates in Al interconnect lines on IC devices [91,92]. By means of the modulated reflectance method it is now possible to detect the presence of a single stacking fault in a Si wafer or a 0.1 μm void within a metal interconnect line. Reviews of semiconductor applications of thermal wave modulated reflectance techniques are provided by Rosencwaig in Ref. [93] and in Chapter 4 of this volume.

IV. LAYERS AND FILMS

Another very important NDE application of thermal waves is the ability to obtain information on layers, films and multilayered samples. This information includes the ability to obtain the thickness of layers and films and the thermal (and sometimes optical) properties of these layers and films. In some cases it is even possible to obtain such information on a multilayered or heterogenous sample.

This capability is related not only to the ability of thermal waves to probe beneath the surface of a sample, but also to the possibility of performing depth profiling as a consequence of the critically damped nature of thermal waves. The concept of thermal wave depth-profiling was first presented in the Rosencwaig-Gersho theory of 1976 [94], demonstrated by Adams and Kirkbright [37] and Afromovitz et al [95] in 1977 and further developed by Rosencwaig in 1978 [96]. Since then depth-profiling, thickness measurements and the analysis of multilayered structures have been addressed both theoretically and experimentally by many authors, some using the frequency dependence of cw thermal waves [97-106], and others the time dependence of transient thermal waves [107-112].

The critically damped nature of thermal waves makes it possible to perform depth-profiling from the front surface of the sample. However, this critical damping also makes tomography, or imaging at fixed depths, extremely difficult with thermal waves, since true tomography requires transmission through the sample. There have, however, been a few interesting experiments demonstrating the possibility of tomography with transmitted thermal waves as a spatially multiplexed cross-sectional imaging technique [113] and as time-of-flight single-shot imaging [See Favro et al, Chapter 2]. If one wishes to study a very thin film (0.1 - 1μm) then, unless one is working with a very small heating spot, it becomes necessary to use very high frequency thermal waves, or very short pulses. A convenient thermal wave technique, employing this principle, is transient thermoreflectance, first developed by Paddock and Eesley [114,115] in 1986 and reviewed by Lorincz and Miklos [116]. Alternatively, one can still employ cw modulated thermal waves at moderate frequencies to measure thin films providing one uses a very small heating spot of 1μm or less, to take advantage of the 3-dimensional aspect of this situation [117]. As shown in Chapter 3 by Opsal and Rosencwaig in this volume, even at quite low frequencies a very thin film will provide a measurable thermal wave signal as long as the heating spot is small.

V. PHYSICAL PROPERTIES

The physical parameters of a material are valuable indicators of the efficacy and repeatability of many important manufacturing processes. Thus there is considerable interest in measuring the optical, electronic, elastic and thermal properties of materials in NDE.

Thermal wave techniques have been very useful for obtaining optical properties of a wide range of materials, especially those that are difficult to measure by conventional optical techniques [118]. Electronic properties of metals [116,117] and of semiconductors [119-122] have also been measured with thermal wave techniques. See also Chapter 5 in this volume by Sheard and Somekh.

Of course the physical properties that are most often measured by thermal wave techniques are the thermal diffusivity and conductivity. Perhaps the first such measurement was that performed by Angstrom in 1863 [123]. Thermal diffusivities and conductivities have been routinely measured since the early days of thermal wave physics [118,124]. Pulsed methods are often used and have been reviewed by Balageas [125]. A photothermal radiometric method utilizing converging rays developed by Cielo *et al* [126] and another method based on optical beam deflection [127,128] developed by Kuo *et al*, both in 1986, have given excellent results. This latter method has proven to be particularly useful for many different samples [129] including thin films, and has been used to measure the thermal conductivity of isotopically enriched diamond single crystals [130] which has the highest known thermal conductivity. Another technique involves a modulated reflectance measurement with displaced pump and probe beams. This technique is particularly useful for very small samples and for materials with anisotopic thermal conductivities, such as high-T_c superconductors [131]. For very thin films, picosecond thermoreflectance has proven to be a successful method since it is not influenced by the thermal properties of the substrate [116,132]. Finally, thermal wave techniques, in general, are particularly useful in determining the thermal properties of powdered and porous materials [103,133,134].

VI. NOVEL TECHNIQUES

As we have seen from this discussion, most thermal wave experiments are performed with optical (usually laser) excitation of the thermal waves. Of course any form of pulsed or modulated excitation can be used, and indeed many different ones have been tried over the years. There are several excitation sources, other than optical, that show some promise for use in thermal wave research and thermal wave NDE. In addition, several novel probes of thermal waves have also been explored. Among the novel excitation and probe radiations that have been tried and appear to be promising are: particle beams such as electrons and ions; x-rays; microwaves; and radio frequency (RF) waves.

(1) Particle Beams

Thermoacoustic signals can be generated by particle beams, and thermoacoustic imaging can be performed in a conventional scanning electron microscope provided the electron beam is modulated. The resultant thermoacoustic signal is usually detected with a piezoelectric transducer in contact with the sample. This was first demonstrated in 1980 by Brandis and Rosencwaig [135] and Cargill [136]. For most materials under normal conditions the dominant mechanism for the generation of the scanning electron acoustic signal is thermoelastic expansion as in the case of laser-ultrasound [137,138]. However, in some materials, particularly semiconductors, mechanisms arising from the injection of electrons into the sample may also play a role [139].

The scanning electron acoustic microscopy method has been used to investigate a variety of materials. Subsurface defects such as voids, cracks and delaminations are readily detected, as are grain boundaries in metals and polycrystalline semiconductors [140-148]. Images of semiconductors exhibit sensitivity to crystalline damage and to the presence of dislocations [149-152]. Ferromagnetic domains can also be imaged [153] as can thermomechanically distinct regions of biological samples [154].

Ion beams can also be used as an excitation source [155-157]. To date, however, only a few experiments have been performed with focused ion beams, and these have addressed the question of the signal generation mechanism in both metals and semiconductors [158,159].

(2) X-Rays

X-rays have long been used in NDE applications, but primarily in the transmission mode. In 1987 Masujima *et al* [160] demonstrated the ability to perform thermal wave studies using x-ray excitation and gas microphone detection. Coufal *et al* [161] have shown that other detection schemes can also be used.

Masujima's group has shown that one can perform x-ray spectroscopy on a variety of materials [162-164], depth profiling and subsurface imaging [165]. See also Chapter 9 by Masujima and Eyring in this volume.

(3) Microwaves

Microwave radiation has long been used to study dielectric and magnetic properties of solids. Netzelmann and Pelzl [166] demonstrated in 1984 that thermal wave techniques can be applied to the detection of microwave absorption in solids that is, where microwaves are used as thermal wave generators. They have also demonstrated that microwaves can be used as the detection probe for thermal waves by modulating the paramagnetic and ferromagnetic resonance through photothermal modulation [167]. This has provided the ability to conduct high spatial resolution probing of magnetic properties [168]. See also Chapter 8 by Pelzl and von Geisau in this volume.

(4) Radio Frequency Waves

RF waves have also found their applications as both a source of excitation and as a probe. Inductive heating has been used as a thermal wave source for years [169] and recently has been combined with photothermal radiometric detection to construct portable thermal wave systems for NDE in the field [170].

In a manner similar to the microwave work described above, photothermal excitation has also been employed to modulate the RF signal in an eddy current detector and thus improve the resolution of eddy current images [171,172].

VII. CONCLUSIONS

Clearly thermal wave techniques have found wide applicability in the NDE area. They are used to generate acoustic and ultrasonic waves, to detect and image subsurface defects, to analyze layers and films and to measure a number of physical properties. Thermal wave methods are already in use in several industries for process development, failure analysis and most importantly for in-line process control and monitoring. As the need for NDE continues to grow in modern manufacturing practices, the importance of thermal wave techniques can also be expected to increase.

VIII. REFERENCES

1. G. Birnbaum and G.S. White, in *Res. Tech. NDT* **7**, (R.S. Sharpe, Ed.) 259 (Academic, New York, 1984).

2. R.M. White, J. Appl. Phys. **34**, 3559 (1963).

3. A. Askar'yan, A.M. Prokhorov, G.F. Chanturiya and M.P. Shipulo, Sov. Phys. JETP **17**, 1463 (1963).

4. R.J. von Gutfeld and R.L. Melcher, Appl. Phys. **30**, 357 (1977).

5. D.A. Hutchins, in *Physical Acoustics* **18**, (W.P. Mason and R.N. Thurston, Eds.) 21 (Academic, New York, 1988).

6. A.C. Tam, in *Photoacoustic and Thermal Wave Phenomena in Semiconductors*, (A. Mandelis, Ed.) 176 (North-Holland, New York,

1982).

7. C.B. Scruby, R.J. Dewhurst, D.A. Hutchins and S.B. Palmer, in *Res. Tech. NDT* **5**, (R.S. Sharpe, Ed.) 281 (Academic, New York, 1982).

8. L.R. Rose, J. Acoust, Soc. Am. **75**, 723 (1984).

9. F.A. McDonald, in *Photoacoustic and Photothermal Phenomena II*, (J.C. Murphy, J.W. Maclachlan Spicer, L.C. Aamodt and B.S.H. Royce, Eds.) 262 (Springer-Verlag, Berlin, 1990).

10. G.A. Alers and H.N.G. Wadley, in *Rev. Prog. Quant. NDE* **6**, (D.O. Thompson and D.E. Chimenti, Eds.) 627 (Plenum, New York, 1987).

11. B.A. Barna, R.T. Allemeier, J.G. Rodriguez and D.M. Tow, in *Rev. Prog. Quant. NDE* **8**, (D.O. Thompson and D.E. Chimenti, Eds.) 1431 (Plenum, New York, 1989).

12. H.I. Ringermacher, B.N. Cassenti, J.R. Strife and J.L. Swindal, in *Rev. Prog. Quant. NDE* **9**, (D.O. Thompson and D.E. Chimenti, Eds.) 471 (Plenum, New York, 1990).

13. B.R. Tittmann, R.S. Lineburger and R.C. Addison, Jr., in *Rev. Prog. Quant. NDE* **8**, (D.O. Thompson and D.E. Chimenti, Eds.) 513 (Plenum, New York, 1989).

14. L.F. Bresse, D.A. Hutchins, F. Hauser and B. Farahbakhsh, in *Rev. Prog. Quant. NDE* **8**, (D.O. Thompson and D.E. Chimenti, Eds.) 527 (Plenum, New York, 1989).

15. A.C. Tam, in *Rev. Prog. Quant. NDE* **8**, (D.O. Thompson and D.E. Chimenti, Eds.), 473 (Plenum, New York, 1989).

16. J.-D. Aussel and J.-P. Monchalin, in *Rev. Prog. Quant. NDE* **8**, (D.O. Thompson and D.E. Chimenti, Eds.) 535 (Plenum, New York, 1989).

17. C.B. Scruby, R.L. Smith and B.C. Moss, NDT Int. **19**, 307 (1986).

18. K. Telschow, in *Rev. Prog. Quant. NDE* **7**, (D.O. Thompson and D.E. Chimenti, Eds.) 1211 (Plenum, New York, 1988).

19. R.J. Dewhurst, C. Edwards, A.D.W. McKie and S.B. Palmer, in *Rev. Prog. Quant. NDE* **7**, (D.O. Thompson and D.E. Chimenti, Eds.) 1615 (Plenum, New York, 1988).

20. J.-P. Monchalin, J.-D. Aussel, P. Bouchard and R. Heon, in *Rev. Prog. Quant. NDE* **7**, (D.O. Thompson and D.E. Chimenti, Eds.) 1607 (Plenum, New York, 1988).

21. L.F. Bressee, D.A. Hutchins and K. Lundgren, in *Rev. Prog. Quant. NDE* **7**, (D.O. Thompson and D.E. Chimenti, Eds.) 1219 (Plenum, New York, 1988).

22. N.M. Carlson and J.A. Johnson, in *Rev. Prog. Quant. NDE* **7**, (D.O. Thompson and D.E. Chimenti, Eds.) 1485 (Plenum, New York, 1988).

23. K.L. Teslchow, J.B. Walter, G.V. Garcia and D.C. Kunerth, in *Rev. Prog. Quant. NDE* **9**, (D.O. Thompson and D.E. Chimenti, Eds.) 2063 (Plenum, New York, 1990).

24. G.V. Garcia, N.M. Carlson, K.L. Telschow and J.A. Johnson, in *Rev. Prog. Quant. NDE* **9**, (D.O. Thompson and D.E. Chimenti, Eds.) 1981 (Plenum, New York, 1990).

25. J.W. Wagner and J.B. Deaton, Jr., in *Rev. Prog. Quant. NDE* **8**, (D.O. Thompson and D.E. Chimenti, Eds.) 505 (Plenum, New York, 1989).

26. R.J. von Gutfeld, D.R. Vigliotti, C.S. Ih and W.R. Scott, Appl. Phys. Lett. **42**, 1018 (1983).

27. R.C. Addison, Jr., L.J. Graham, R.S. Lineburger and B.R. Tittman, in *Rev. Prog. Quant. NDE* **7**, (D.O. Thompson and D.E. Chimenti, Eds.) 585 (Plenum, New York, 1988).

28. J.-P. Monchalin, IEEE Trans. UFFC-**33**, 485 (1986).

29. P. Charpentier, F. Lepoutre and L. Bertrand, J. Appl. Phys. **53**, 608 (1982).

30. P. Cielo, X. Maldague, G. Rousset and C.K. Jen, Mat. Eval. **43**, 1111 (1983).

31. K. Hane, T. Kanie and S. Hattori, J. Appl. Phys. **64**, 2229 (1988).

32. K. Hane and S. Hattori, Appl. Opt. **27**, 3965 (1988).

33. K. Hane, T. Naito and S. Hattori, Appl. Opt. **30**, 72 (1991).

34. A. Rosencwaig, Opt. Commun. **7**, 305 (1973).

35. J.G. Parker, J. Chem. Phys. **36**, 1547 (1962); and Appl. Opt. **12**, 2974 (1973).

36. *Optoacoustic Spectroscopy and Detection*, (Y.-H. Pao, Ed.) (Academic, New York, 1977).

37. M.J. Adams and G.F. Kirkbright, Analyst, **102**, 678 (1977).

38. Y.H. Wong, R.L. Thomas and G.F. Hawkins, Appl. Phys. Lett. **32**, 538 (1978).

39. G. Busse, Appl. Phys. Lett. **35**, 759 (1979).

40. M. Luukkala and A. Penttinen, Electron. Lett. **15**, 326 (1979).

41. R.L. Thomas, J.J. Pouch, Y.H. Wong, L.D. Favro, P.K. Kuo and A. Rosencwaig, J. Appl. Phys. **51**, 1152 (1980).

42. A. Rosencwaig and G. Busse, Appl. Phys. Lett. **36**, 725 (1980).

43. S.-Y. Zhang and L. Chen, in *Photoacoustic and Thermal Wave Phenomena in Semiconductors*, (A. Mandelis, Ed.) 27 (North-Holland, New York, 1987).

44. K.R. Grice, L.J. Inglehart, L.D. Favro, P.K. Kuo and R.L. Thomas, J. Appl. Phys. **54**, 6245 (1983).

45. A.C. Tam and H. Coufal, Appl. Phys. Lett. **42**, 33 (1983).

46. J.F. McClelland and R.N. Kniseley, Appl. Phys. Lett. **35**, 121 (1979).

47. M. Kasai, H. Shimizu, T. Sawada and Y. Gohshi, Anal. Sci. **1**, 107 (1985).

48. M. Kasai, T. Sawada, Y. Gohshi, T. Watanabe and K. Furuya, Jpn. J. Appl. Phys. Suppl. **25-1**, 229 (1986).

49. A.C. Boccara, D. Fournier and J. Badoz, Appl. Phys. Lett. **36**, 130 (1980).

50. J.C. Murphy and L.C. Aamodt, J. Appl. Phys. **51**, 4580 (1980).

51. J.C. Murphy and L.C. Aamodt, J. Appl. Phys. **54**, 581 (1983).

52. M.J. Lin, L.J. Inglehart, L.D. Favro, P.K. Kuo and R.L. Thomas, in *Rev. Prog. Quant. NDE* **4**, (D.O. Thompson and D.E. Chimenti, Eds.) 739 (Plenum, New York, 1985).

53. V.P. Vavilov and R. Taylor, in *Res. Tech. NDT* **5**, (R.S. Sharpe, Ed.) (Academic, New York, 1982).

54. W.N. Reynolds, Can. J. Phys. **64**, 1150 (1986).

55. P.-E. Nordal and S.O. Kanstad, in *Scanned Image Microscopy*, (E.A. Ash, Ed.) 331 (Academic, London, 1980).

56. G. Busse, in *Scanned Image Microscopy* (E.A. Ash, Ed.) 341 (Academic, London, 1980).

57. S.O. Kanstad and P.-E. Nordal, Can. J. Phys. **64**, 1155 (1986).

58. A. Lehto, J. Jaarinen, T. Tiusanen, M. Jokinen and M. Luukkala, Electron. Lett. **17**, 364 (1981).

59. D.P. Almond, P.M. Patel and H. Reiter, J. Phys. (Paris) **C6-44**, 491 (1983).

60. P. Cielo, J. Appl. Phys. **56**, 230 (1984).

61. H. Ermest. F.J. Dacol, R.L. Melcher and T. Baumann, Appl. Phys. Lett. **44**, 1136 (1984).

62. G. Busse, Infrared Phys. **20**, 419 (1980).

63. M. Luukkala, in *Scanned Image Microscopy*, (E.A. Ash, Ed.) 273 (Academic, London, 1980).

64. G. Busse, Appl. Opt. **21**, 109 (1982).

65. G. Busse and K.F. Renk, Appl. Phys. Lett. **42**, 366 (1983).

66. G. Busse and P. Eyerer, Appl. Phys. Lett. **43**, 355 (1983).

67. P.K. Kuo, Z.J. Feng, T. Ahmed, L.D. Favro, R.L. Thomas and J. Hartikainen, in *Photoacoustic and Photothermal Phenomena*, (P. Hess and J. Pelzl, Eds.) 415 (Springer-Verlag, Berlin, 1988).

68. T. Ahmed, P.K. Kuo, L.D. Favro, H.J. Jin and R.L. Thomas, in *Rev.*

Prog. Quant. NDE **8**, (D.O. Thompson and D.E. Chimenti, Eds.) 607 (Plenum, New York, 1989).

69. T. Ahmed, H.J. Jin, P. Chen, P.K. Kuo, L.D. Favro and R.L. Thomas, in *Photoacoustic and Photothermal Phenomena II*, (J.C. Murphy, J.W. Maclachlan Spicer, L.C. Aamodt and B.S.H. Royce, Eds.) 30 (Springer-Verlag, Berlin, 1990).

70. V. Vavilov, T. Ahmed, H.J. Jin, R.L. Thomas and L.D. Favro, Sov. J. NDT **12**, 65 (1990).

71. J.W.M. Spicer, W.D. Kerns, L.C. Aamodt and J.C. Murphy, J. Nondestructive Eval. **8**, 107 (1989).

72. L.C. Aamodt, J.W.M. Spicer and J.C. Murphy, J. Appl. Phys. **68**, 6087 (1990).

73. A.A. Deom, D. Boscher and D.L. Balageas, in *Rev. Prog. Quant. NDE* **9**, (D.O. Thompson and D.E. Chimenti, Eds.) 525 (Plenum, New York, 1990).

74. J. Hartikainen, Rev. Sci. Instr. **60**, 670 and 1334 (1989).

75. V. Thien, P. Vetterlein, M. Kroning and E. Winschuh, Microchim. Acta. **II**, 25 (1990).

76. S. Ameri, E.A. Ash, V. Neuman and C.R. Petts, Electron Lett. **17**, 337 (1981).

77. H.K. Wickramasinghe, Y. Martin, D.A. H. Spear and E.A. Ash, J. Phys. (Paris) **C6-44**, 191 (1983).

78. N.M. Amer, J. Phys. (Paris) **C6-44**, 185 (1983).

79. J. Opsal, A. Rosencwaig and D.L. Willenborg, Appl. Opt. **22**, 3169 (1983).

80. A. Rosencwaig, J. Opsal, W.L. Smith and D.L. Willenborg, Appl. Phys. Lett. **46**, 1013 (1985).

81. J. Opsal and A. Rosencwaig, Appl. Phys. Lett. **47**, 498 (1985).

82. W.L. Smith, A. Rosencwaig and D.L. Willenborg, Appl. Phys. Lett. **47**, 584 (1985).

83. J. Opsal, M.W. Taylor, W.L. Smith and A. Rosencwaig, J. Appl. Phys. **61**, 240 (1986).

84. L.A. Vitkin, C. Christofides and A. Mandelis, Appl. Phys. Lett. **54**, 2392 (1989).

85. I.-W.H. Connick, A. Bhattacharyva, K.N. Ritz and W.L. Smith, J. Appl. Phys. **64**, 2059 (1988).

86. A. Rosencwaig, J. Opsal, D.L. Willenborg, P. Geraghty and W.L. Smith, in *Photoacoustic and Photothermal Phenomena*, (P. Hess and J. Pelzl, Eds.) 229 (Springer-Verlag, Berlin, 1988).

87. B. Witowski, W.L. Smith and D.L. Willenborg, Appl. Phys. Lett. **52**, 640 (1988).

88. P. Alpern, W. Bergholz and R. Kokoschke, J. Electrochem. Soc. **136**, 3841 (1989).

89. S. Hahn, W.L. Smith, H. Suga, R. Meinecke, R.R. Kola and G.A. Rozgonyi, J. Cryst. Growth **103**, 206 (1990).

90. W.L. Smith, C.G. Welles and A. Rosencwaig, in *Rev. Prog. Quant NDE* **9**, (D.O. Thompson and D.E. Chimenti, Eds.) 1087 (Plenum, New York, 1990).

91. W.L. Smith, C.G. Welles, D.L. Willenborg and A. Rosencwaig, in *Tech. Proc. Semicon Japan Osaka 1989*, SEMI, Santa Clara, 1989.

92. W.L. Smith, C.G. Welles, A. Bivas, F.G. Yost and J.E. Campbell, *Proc. 1990 IRPS, IEEE*, 200, New York, 1990.

93. A. Rosencwaig, in *Photoacoustic and Thermal Wave Phenomena in Semiconductors*, (A. Mandelis, Ed.) 97 (North-Holland, New York, 1987).

94. A. Rosencwaig and A. Gersho, J. Appl. Phys. **47**, 64 (1976).

95. M.A. Afromovitz, P.S. Yeh and S.S. Yee, J. Appl. Phys. **48**, 209 (1977).

96. A. Rosencwaig, J. Appl. Phys. **49**, 2905 (1978).

97. A. Rosencwaig, in *Adv. Electron. Electron Phys.* **46**, (L. Marton, Ed.)

208 (Academic, New York, 1978).

98. P. Korpiun and R. Tilgner, J. Appl. Phys. **51**, 6115 (1981).

99. J. Opsal and A. Rosencwaig, J. Appl. Phys. **53**, 4240 (1982).

100. W. Goertz and H.H. Perkampus, Anal. Chem. **310**, 77 (1982).

101. D. Fournier, F. Lepoutre and A.C. Boccara, J. Phys. (Paris) **C6-44**, 479 (1983).

102. D.M. Anjo and T.A. Moore, Photochem. Photobiol. **39**, 635 (1984).

103. A. Mandelis and J.D. Lymer, Appl. Spectrosc. **39**, 473 (1985).

104. P. Cielo, G. Rousset and L. Bertrand, Appl. Opt. **25**, 1327 (1986).

105. A. Harata and T. Sawada, J. Appl. Phys. **65**, 959 (1989).

106. A. Mandelis, E. Schoubs, S.B. Peralta and J. Thoen, J. Appl. Phys. **70**, 1771 (1991).

107. A.C: Tam, Rev. Mod. Phys. **58**, 38 (1981).

108. A. Uejima, F. Itoga and Y. Sugitani, Anal. Sci. **2**, 113 (1986).

109. J.W. Maclachlan Spicer, W.D. Kerns, L.C. Aamodt and J.C. Murphy, in *Photoacoustic and Photothermal Phenomena II*, (J.C. Murphy, J.W. Maclachlan Spicer, L.C. Aamodt and B.S.H. Royce, Eds.) 55 (Springer-Verlag, Berlin, 1990).

110. L.C. Aamodt, J.W. Maclachlan Spicer and J.C. Murphy, in *Photoacoustic and Photothermal Phenomena II*, (J.C. Murphy, J.W. Maclachlan Spicer, L.C. Aamodt and B.S.H. Royce, Eds.) 59 (Springer-Verlag, Berlin, 1990).

111. V.P. Vavilov, SPIE **1313**, Thermosense XII, 178 (1990).

112. L.D. Favro, T. Ahmed, D. Crowther, H.J. Jin, P.K. Kuo, R.L. Thomas and X. Wang, SPIE **1467**, Thermosense XIII, 132 and 290 (1991).

113. M. Munidasa and A. Mandelis, J. Opt. Soc. Am. A. **8**, 185 (1991); A. Mandelis, J. Phys. A: Math. Gen. **24**, 2485 (1991).

114. C.A. Paddock and G.L. Eesley, Opt. Lett. **11**, 273 (1986).

115. C.A. Paddock and G.L. Eesley, J. Appl. Phys. **60**, 285 (1986).

116. A. Lorincz and A. Miklos, in *Principles and Perspectives of Photothermal and Photoacoustic Phenomena*, (A. Mandelis, Ed.) 431 (Elsevier, New York, 1992).

117. A. Rosencwaig, in *Rev. Prog. Quant. NDE* **9**, (D.O. Thompson and D.E. Chimenti, Eds.) 2031 (Plenum, New York, 1990).

118. A. Rosencwaig, *Photoacoustics and Photoacoustic Spectroscopy*, J. Wiley, New York, 1980.

119. D. Fournier and A.C. Boccara, in *Photoacoustic and Thermal Wave Phenomena in Semiconductors*, (A. Mandelis, Ed.) 237 (North-Holland, New York, 1987).

120. A.C. Boccara and D. Fournier, in *Photoacoustic and Thermal Wave Phenomena in Semiconductors*, (A. Mandelis, Ed.) 287 (North-Holland, New York, 1987).

121. S.J. Sheard, *Proc. IEEE Ultrasonic Symp.*, 789 (1986).

122. S.J. Sheard and M.G. Somekh, Infrared Phys. **28**, 287 (1988).

123. M.A.J. Angstrom, Phil. Mag. **25**, 130 (1863).

124. H. Vargas and L.C.M. Miranda, Physics Reports **161**, 45 (1988).

125. D.L. Balageas, High Temp.-High Press. **21**, 85 (1989).

126. P. Cielo, L.A. Ultracki and M. Lamontagne, Can. J. Phys. **64**, 1172 (1986).

127. P.K. Kuo, M.J. Lin, C.B. Reyes, L.D. Favro, R.L. Thomas, D.S. Kim, S.-Y. Zhang, L.J. Inglehart, D. Fournier, A.C. Boccara and N. Jacoubi, Can. J. Phys. **64**, 1165 (1986).

128. P.K. Kuo, E.D. Sendler, L.D. Favro and R.L. Thomas, Can. J. Phys. **64**, 1168 (1986).

129. S. Salazar, A. Sanchez-Lavega and J. Fernandez, J. Appl. Phys. **69**, 1216 (1991).

130. T.R. Anthony, W.F. Banholzer, J.F. Fleischer, L. Wei, P.K. Kuo and R.W. Pryor, Phys. Rev. B **42**, 1104 (1990).

131. J.T. Fanton, A. Kapitulnik, D.B. Mitzi, B.T. Khuri-Yakub and G.S. Kino, in *Photoacoustic and Photothermal Phenomena II*, (J.C. Murphy, J.W. Maclachlan Spicer, L.C. Aamodt and B.S.H. Royce, Eds.) 220 (Springer-Verlag, Berlin, 1990).

132. A. Miklos and A. Lorincz, Appl. Phys. B **48**, 261 (1989); O.B. Wright, T. Hyoguchi and K. Kawashima, Jpn. J. Appl. Phys. **30**, L131 (1991).

133. A. Boccara and D. Fournier, *Photoacoustic and Photothermal Phenomena*, (P. Hess and J. Pelzl, Eds.) 302 (Springer-Verlag, Berlin, 1988).

134. R. Osiander, R. Haberkern, P. Korpiun and W. Schirmacher, in *Photoacoustic and Photothermal Phenomena II*, (J.C. Murphy, J.W. Maclachlan Spicer, L.C. Aamodt and B.S.H. Royce, Eds.) 309 (Springer-Verlag, Berlin, 1990).

135. E. Brandis and A. Rosencwaig, Appl. Phys. Lett. **37**, 98 (1980).

136. G.S. Cargill, III, Nature **286**, 691 (1980).

137. A. Rosencwaig and J. Opsal, IEEE UFFC-**33**, 516 (1986).

138. J.W. Maclachlan and J.C. Murphy, in *Photoacoustic and Photothermal Phenomena*, (P. Hess and J. Pelzl, Eds.) 294 (Springer-Verlag, Berlin, 1988).

139. L.J. Balk, Can. J. Phys. **64**, 1238 (1986).

140. A. Rosencwaig, Science **218**, 223, (1982).

141. A. Rosencwaig, Solid State Technology, Mar. 1982, p. 91.

142. A. Rosencwaig, J. Scan. Electron Microsc. **4**, 1611 (1984).

143. A. Rosencwaig, Ann. Rev. Mater. Sci. **15**, 103 (1985).

144. D.G. Davies, J. Scan. Electron Microsc. **3**, 1163 (1983).

145. G.S. Cargill, III, in *Opto- and Photoacoustic Spectroscopy*, (J.F.

McClelland and C.K.N. Patel, Eds.) (Academic, New York, 1984).

146. L.J. Balk, in *Adv. Electron. and Electron Physics* **71**, (L. Marton, Ed.) 1 (Academic, New York, 1988).

147. F. Dominguez-Adams and J. Piqueras, J. Appl. Phys. **66**, 2751 (1989).

148. J.H. Cantrell, M. Qian, M.V. Ravichandran and K.M. Knowles, Appl. Phys. Lett. **57**, 1870 (1990).

149. A. Rosencwaig and R.M. White, Appl. Phys. Lett. **38**, 165 (1981).

150. L.J. Balk and N. Kultscher, J. Phys. **C2**, 873 (1984).

151. T.D. Kirkendall and T.P. Remmel, J. Phys. C **2**, 877 (1984).

152. H. Takenoshita, Jpn. J. Appl. Phys. Suppl. **30-1**, 253 (1991).

153. L.J. Balk, D.G. Davies and N. Kultscher, IEEE Trans. Magn. **MAG-20**, 1466 (1984).

154. L.J. Balk, D. Domnik, K. Niklas and P. Mestres, in *Photoacoustic and Photothermal Phenomena II*, (J.C. Murphy, J.W. Maclachlan Spicer, L.C. Aamodt and B.S.H. Royce, Eds.) 428 (Springer-Verlag, Berlin, 1990).

155. D.N. Rose, H. Turner and K.O. Legg, Can. J. Phys. **64**, 1284 (1986).

156. K. Kimura, K. Nakanishi, A. Nishimura and M. Mannami, Jpn. J. Appl. Phys. **24**, L449 (1985).

157. F.G. Satkiewicz, J.C. Murphy, L.C. Aamodt and J.W. Maclachlan, in *Rev. Prog. Quant. NDE* **5**, (D.O. Thompson and D.E. Chementi, Eds.) 789 (Plenum, New York, 1986).

158. F.G. Satkiewicz, J.C. Murphy, J.W. Maclachlan and L.C. Aamodt, in *Rev. Prog. Quant. NDE* **6**, (D.O. Thompson and D.E. Chimenti, Eds.) 759 (Plenum, New York, 1987).

159. F.G. Satkiewicz, J.C. Murphy and L.C. Aamodt, in *Photoacoustic and Photothermal Phenomena*, (P. Hess and J. Pelzl, Eds.) 288 (Springer-Verlag, Berlin, 1988).

160. T. Masujima, H. Kawata, Y. Amemiya, N. Kamuja, T. Katsura, T.

Iwamoto, H. Yoshida, H. Imai and M. Ando, Chem. Lett. **5**, 973 (1987).

161. H. Coufal, J. Stohr and K. Baberschke, in *Photoacoustic and Phothermal Phenomena*, (P. Hess and J. Pelzl, Eds.) 25 (Springer-Verlag, Berlin, 1988).

162. T. Masujima, H. Yoshida, H. Katawa, Y. Amemiya, T. Katsura, M. Ando, K. Fukui and M. Watanabe, Rev. Sci. Instrum. **60**, 2318 and 2522 (1989).

163. T. Masujima, H. Shiwaku, H. Yoshida, M. Kataoke, M. Reichling, H. Imai, H. Kawata, A. Iida, A. Koyama and M. Ando, Jpn. J. Appl. Phys. **28**, L513 (1989).

164. T. Toyoda, T.Masujima, H. Schiwaku, A. Iida and M. Ando, Jpn. J. Appl. Phys. **29**, L1723 and 2541 (1990).

165. T. Masujima, in *Photoacoustic and Photothermal Phenomena II*, (J.C. Murphy, J.W. Maclachlan Spicer, L.C. Aamodt and B.S.H. Royce, Eds.) 222 (Springer-Verlag, Berlin, 1990).

166. U. Netzelmann and J. Pelzl, Appl. Phys. Lett. **44** 854 and 1161 (1984).

167. T. Orth, U. Netzelmann and J. Pelzl, Appl. Phys. Lett. **53**, 1979 (1988).

168. J. Pelzl, U. Netzelmann, T. Orth and R. Kordecki, in *Photoacoustic and Photothermal Phenomena II*, (J.C. Murphy, J.W. Maclachlan Spicer, L.C. Aamodt and B.S.H. Royce, Eds.) 2 (Springer-Verlag, Berlin, 1990).

169. J. Saniie, M. Luukkala, A. Letho and R. Rajala, Electron, Lett. **18**, 651 (1982).

170. R. Lehtiniemi, J. Hartikainen, J. Varis and M. Luukkala, in *Photoacoustic and Photothermal Phenomena III*, (D. Bicanic, Ed.) 512 (Springer-Verlag, Berlin, 1992).

171. J.C. Moulder, M.W. Kubovich, J.M. Mann, M.S. Hughes and N. Nakagawa, in *Rev. Prog. Quant. NDE* **9**, (D.O.Thompson and D.E. Chimenti, Eds.) 533 (Plenum, New York, 1990).

172. J.C. Moulder, D.N. Rose, D.C. Bryk and J.S. Siwicki, in *Rev. Prog. Quant. NDE* **9**, (D.O. Thompson and D.E. Chimenti, Eds.) 539 (Plenum, New York, 1990).

REAL-TIME THERMAL WAVE IMAGING

L.D. Favro, P.K. Kuo and R.L. Thomas

Institute for Manufacturing Research and Department of Physics
Wayne State University, Detroit, MI 48202

I.	INTRODUCTION	24
	1. Sinusoidal Heat Source	24
	2. Pulsed Heat Source	25
	3. Moving Heat Source	27
II.	BOX-CAR THERMAL WAVE IMAGING	29
III.	PULSED THERMAL-WAVE TOMOGRAPHY	35
IV.	INVERSE SCATTERING AND OBJECT RECONSTRUCTION	38
V.	LOCK-IN VIDEO THERMAL WAVE IMAGING	40
VI.	THE FLYING SPOT CAMERA	42
VII.	REFERENCES	47

I. INTRODUCTION

Real-time thermal wave imaging exploits the time-dependent heat flow that takes place when a time varying heat source is applied to an object. These heat sources usually apply the heat either in the form of a long train of more or less sinusoidal pulses with equal on/off periods, or, alternatively, in the form of a short pulse followed by a relatively long period with no heating. The result is the launching of "thermal waves", i.e. waves of temperature variation into the object [1]. As might be expected, these waves are periodic when the source is periodic and transient when the source is pulsed. A third, less common, source of thermal waves is a rapidly moving laser beam focussed on the surface of a material. This type of source leaves a time-dependent "thermal wake" behind it, and can also be used to make thermal wave images [2]. In this section we will explore the behavior of these three types of waves in homogeneous media. This will serve as a basis for the discussion of the imaging mechanisms described in later sections.

1. Sinusoidal Heat Source

Time dependent heat flow is governed by the heat diffusion equation,

$$\kappa \nabla^2 T = \rho c \frac{\partial T}{\partial t} . \qquad (1)$$

In this equation, T represents the temperature excursion from the equilibrium or ambient temperature, and κ, ρ, and c respectively represent the thermal conductivity, the density, and the specific heat of the medium. These last three material constants are often combined into a single material parameter called the thermal diffusivity, α, which is defined as $(\kappa/\rho c)$. When the temperature excursion described by Eq. (1) is generated by a periodic source whose periodicity is represented by the complex form $\exp(-i\omega t)$, this equation reduces to the well-known Helmholtz wave equation,

$$\nabla^2 T + q^2 T = 0 , \qquad (2)$$

where q is the (complex) wave number of the periodic thermal wave,

$$q = \sqrt{\frac{i\omega}{\alpha}} = (1+i)\sqrt{\frac{\omega}{2\alpha}} . \qquad (3)$$

A simple one-dimensional solution to this equation has the form

$$T(x) = T_o e^{i(qx-\omega t)} \quad . \tag{4}$$

It is the real part of this expression which represents the actual physical temperature excursion,

$$T(x) = T_o \cos\left(\sqrt{\frac{\omega}{2\alpha}}\, x - \omega t\right) \exp\left(-\sqrt{\frac{\omega}{2\alpha}}\, x\right) . \tag{5}$$

Thus it is seen that a periodic thermal wave is the product of a travelling wave having a strongly frequency dependent velocity, $\upsilon = \sqrt{2\alpha\omega}$, with a strong exponential damping term which has a decay length (thermal diffusion length) given by $\mu = \sqrt{2\alpha/\omega}$. These two factors cause thermal waves to be very dispersive, and to decay very rapidly as they propagate into the material. In fact, the amplitude of a thermal wave decreases by a factor of 500 in a distance of just one wavelength. Nonetheless, these waves can be scattered or reflected by thermal inhomogeneities, and can be used to make images of near-subsurface thermal features in an object.

2. Pulsed Heat Source

In the description above, the thermal waves were imagined to have been launched by a periodic heat source, for example, a modulated laser beam. Another common source of heat in thermal wave experiments is a xenon flash lamp, or an array of such lamps. In this case, the heat is applied in a short pulse, which may be idealized as a pulse of infinitely short duration, but with a finite amount of heat deposition. The appropriate one-dimensional solution to Eq. 1 then is of the form,

$$T(x,t) = \frac{C}{\sqrt{4\pi\alpha t}}\, e^{-\frac{x^2}{4\alpha t}} , \tag{6}$$

where C is a constant which is related to the amount of heat deposited on the sample surface (assumed to be located at $x = 0$). At $t = 0$, this pulse is infinitely narrow and located at the surface $x = 0$. As the pulse propagates into the sample, the strong frequency dependence of the velocity described above causes the pulse to spread very rapidly, and the strong attenuation causes its amplitude to decrease very rapidly. In Figure 1 we show plots of the pulse shape as a function of time for representative depths beneath the surface. The time of arrival of the peak of the pulse, and the amplitude of the peak temperature at any given depth x are given, respectively, by

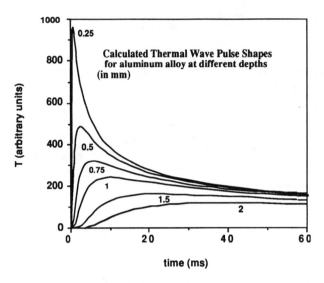

Fig. 1 A plot of a thermal wave pulse as a function of time at several depths beneath the surface. The units are chosen such that $\alpha = 0.5$.

$$t_{peak} = \frac{x^2}{2\alpha}, \qquad (7)$$

and

$$T_{peak} = \frac{C}{\sqrt{2\pi e}} \frac{1}{x}. \qquad (8)$$

The long "tails" on the pulses shown in Fig. 1 result from the fact that the low frequency components of the thermal wave pulse propagate with very low velocities and hence take a long time to cross any given point under the surface. Pulses such as those shown in Fig. 1 can be reflected or scattered by subsurface thermal inhomogeneities and return to the surface of the sample as greatly broadened and attenuated "echoes" of the initial sharp pulse. With proper instrumentation, these echoes may be used to form images of the subsurface features from which they were scattered. It is instructive to evaluate the arrival time of the peak of a thermal wave pulse for a typical industrial sample. For instance, the skin on the fuselage of a commercial aircraft might consist of a 40 mil (1mm) sheet of rolled aluminum alloy with a thermal diffusivity of the order of 0.5 cm²/s. The time for the peak of the pulse to penetrate this skin, be reflected and return to the surface is then given by

$$t_{peak} = \frac{(0.2\,cm)^2}{2(0.5\,cm^2/s)} = 40\,ms$$

It is noteworthy to recognize that this time is on the same order of magnitude as the frame time, 33 ms, of an NTSC video camera. This means that when using infrared video cameras to form images, one often must operate at speeds which are faster than video frame rates to accurately record thermal wave effects in thin specimens with high thermal diffusivities. Methods which can accomplish this will be described in the sections which follow.

3. Moving Heat Source

When the heat source is a moving laser beam it is no longer possible to describe the heat flow in one dimension. Instead, one must use the three-dimensional equivalent of Eq. 6, and integrate it over the path of the beam from its starting position to the present position. If the beam's path is a long straight line, this integral has the form

$$T(r,t) = \int_{-\infty}^{t} \frac{C}{[4\pi\alpha(t-t')]^{3/2}} e^{-\frac{(r-vt')^2}{4\alpha(t-t')}} dt' \quad (9)$$

where the vector **v** represents the velocity of the beam on the surface. It is possible to perform this integral analytically to obtain an explicit expression for the temperature,

$$T(r,t) = \frac{C}{4\pi\alpha} \cdot \frac{1}{[y^2+z^2+(x-vt)^2]^{1/2}} \exp\left(-\frac{v}{2\alpha}\{(x-vt) + [y^2+z^2+(x-vt)^2]^{1/2}\}\right) \quad (10)$$

where we have assumed that the surface of the sample is the x-y plane and that the beam is moving along the x-axis. The temperature distribution along the track of the beam calculated from Eq. (10) is plotted in Fig. 2 for a diffusivity of 0.25 cm^2/s, and for values of velocity ranging from 1 m/s to 100 m/s.

The temperature distribution shown in Fig. 2 has a long thermal "wake" which varies inversely with distance behind the heated laser spot, and an exponential precursor which projects out in front of the spot. The wake just represents the cooling of the previously heated track, and is independent of velocity. The reason for this surprising result stems from the cancellation of two effects: first, the dwell time of the laser spot, and hence the amount of heat deposited, is inversely proportional to the velocity; second, the delay

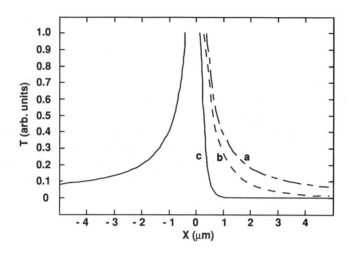

Fig. 2 A plot of surface temperature vs. distance along the track of a moving laser spot on the surface of an opaque sample for a diffusivity of 0.25 cm²/s, and for values of velocity: a) 1m/s, b) 10 m/s, and c) 100 m/s.

from the application of the heat for a fixed distance behind the laser spot is also inversely proportional to the velocity. Since the heat is effectively deposited in a line, the decay rate of the temperature is proportional to the reciprocal of the delay time, and thus a shorter delay time exactly compensates for a shorter dwell time. This wake consists mostly of the slow, low-frequency components of the heat pulse which take a long time to leave the surface. The precursor results from the propagation of thermal waves into the region ahead of the moving beam. It is rich in the fast, high-frequency, thermal wave components generated by the laser beam. As the velocity increases, the precursor is compressed toward the source, as one might expect. In Figure 3, we plot the surface temperature versus distance along the track of a moving laser spot on the surface of an opaque sample for a velocity of 10 m/s, and for values of diffusivities ranging from 0.1 cm²/s to 0.5 cm²/s. It may be noted from Fig. 3 that the amplitudes of the thermal wakes vary inversely with the thermal diffusivities. The precursor is also inversely proportional to the thermal diffusivity near the source. However, farther ahead of the source, the slower exponential decay causes the precursors to merge and eventually cross. Just as in the cases of waves generated by sinusoidal or pulsed sources, the thermal waves generated by a moving spot can also be used to image subsurface thermal features of a sample. Examples of all these applications will be given in the sections which follow.

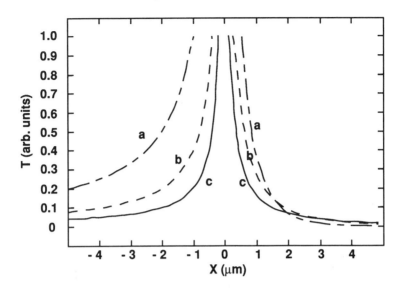

Fig. 3 A plot of surface temperature versus distance along the track of a moving laser spot on the surface of an opaque sample for a velocity of 10 m/s, and for values of diffusivities: a) 0.1 cm²/s, b) 0.25 cm²/s, and c) 0.5 cm²/s.

II. BOX-CAR THERMAL WAVE IMAGING

When imaging a subsurface defect with a heat source which is pulsed [3-46], it is often advantageous to form an image of the surface temperature distribution corresponding to the time when the peak of the thermal wave echo from the defect has just returned to the surface. A thermal wave image made at this time will exhibit the maximum contrast between the defect region and the background. To achieve this end it is desirable to have an imaging system which is capable of making gated images at some fixed time after the occurrence of the heating pulse. An imaging system which operates in this fashion is analogous to the gated (single channel) signal acquisition system which is commonly referred to as a "box-car integrator". A block diagram of the Wayne State Box-Car Imaging System is shown in Fig. 4. This system is controlled by a computer (Sun workstation, Macintosh II, or IBM 486 clone) which is used to synchronize the components of the system, download the software for the real-time processor, and to display and post-process the resulting thermal wave images. The heat source consists of a battery of up to eight 6.4 kJ, 2 ms pulse duration, xenon flash lamps, which are fired simultaneously by a signal from the computer. The surface temperature of the sample is monitored as a function of time by an infrared video camera (Inframetrics Model 600), operating in the 8µm to 12µm region of the IR. The stream of analog data coming from the camera is digitized in the real-time processor (DataCube, Perceptics), which is programmed to carry out the boxcar gating in real time. In this mode of operation of the system, an image, or series of images, is acquired at fixed times after the occurrence of the flash. The intention is to capture the returning thermal wave echoes from an subsurface defects when their constrast with the background is near its maximum value.

The operation of this system can be illustrated by describing its application to the box-car imaging of a test specimen (Fig. 5) in which a series of flat-bottomed holes have been milled into the back surface at incrementally increasing depths beneath the front surface.

The interface between the solid and air at the flat subsurface boundary of each hole serves as a convenient subsurface thermal wave reflector to produce an echo pulse. If we imagine that this sample is of a new and unknown material, the first step is to determine where to set the gates of the imager to capture the echoes. To this end, an exploratory pulse is sent into the sample by flashing the lamps once and visually observing the returning pulses from any defects which may be present through the use of the IR camera's monitor screen. In the case of our flat-bottomed hole sample, one will see a set of more or less distinct circular patches which appear and then fade away. Then, using the software package in the control computer and a

Fig. 4 Block diagram of the Box-car Imaging System at Wayne State University.

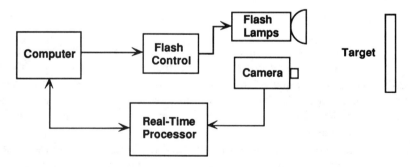

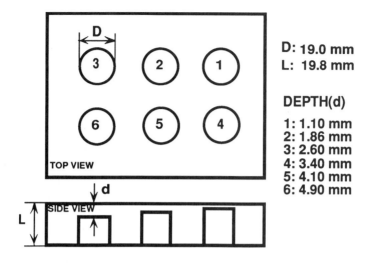

Fig. 5 Schematic diagram of the flat-bottomed hole test specimen.

Fig. 6 Experimental heating and cooling curves for a plastic sample with six flat-bottomed holes milled into the back surface. The lowest curve is a reference taken over a region with no subsurface defects. The other six curves are from areas directly over the six holes.

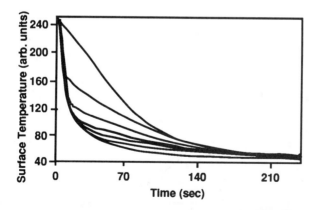

mouse attached to it, one would define seven small rectangular areas, six over the regions where the reflections had been seen, and one elsewhere for reference. The real time processor is then instructed to spatially average the temperature in each test rectangle and plot each average as a function of time after the next flash. Upon flashing the lamps again, a set of curves such as those shown in Fig. 6 appears on the screen of the computer's monitor. These curves show the heating and cooling of the surface of the sample in each of the selected regions. The differences in the cooling curves are due to the returning thermal wave echoes from the buried holes. Since the holes are all at different depths beneath the heated surface, both the amplitudes and the peak times of the echoes are different, making each curve distinct. The time evolution of the echo pulses can be seen in Fig. 7 where the background time dependence has been removed by subtraction of the reference curve from each of the other six. It can be seen that echoes with larger amplitudes peak earlier as predicted for shallower reflectors by the theory above. These observations can be made more quantitative by plotting the peak times versus the square of the hole's depth to determine the validity of Eq. (7) for the reflected pulses. A plot of this type corresponding to the curves of Fig. 7, is shown in Fig. 8. It is apparent from the plot that the relationship between the two variables is linear.

The determination of the peak times for the returning echoes need not be repeated when one is working with a series of similar samples, or when working with samples which have been previously characterized either by measurements or by theoretical calculations. Having determined peak times

Fig. 7. Plots of the returning thermal wave echoes from the six holes in the sample of Fig. 6. These are obtained by subtracting the reference curve in Fig. 6 from the remaining six curves. Note the similarity between these experimental curves and the theoretical curves of Fig. 1.

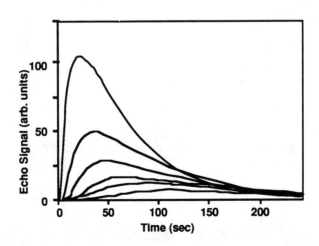

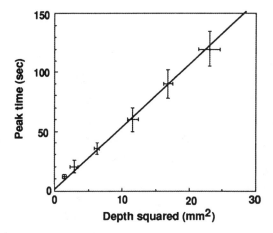

Fig. 8 A plot of the peak times vs. the square of the depth beneath the heated surface for the echo pulses shown in Fig. 7.

for the echoes from various kinds of defects of interest in a sample, one can proceed to make actual box-car images. In these images the real-time processor is set to acquire a time-gated image in a window which includes the peaks of the echoes of interest, and also to acquire a second image in a gated window set relatively far out on the tails of the echo curves. The function of this second image is to serve as a reference image which, when it is subtracted from the image acquired in the first gate, tends to remove many of the background artifacts from the combined image. The resulting combined image may be averaged over several flashes if it is deemed necessary to improve the signal-to-noise ratio. Other arithmetic operations, such as division of one image by another, can also be performed by the real-time processor. An example of a (subtracted) box-car image of the flat-bottomed hole sample from which the curves above were taken is shown in Fig. 9. This image was made with the first gate set at 0.75 s. In this image lighter gray scale indicates larger signal strength. Notice that the deeper holes appear less distinct because of the smaller amplitudes of their thermal wave echoes. Often one finds that it is not possible to select a gate time which enables all of the defects in a sample to be shown in the same image. This happens when the variation in the depths of the defects is so large that the earliest arriving echoes have faded before the deeper ones have peaked. In this case it is necessary to make two or more images with different gate

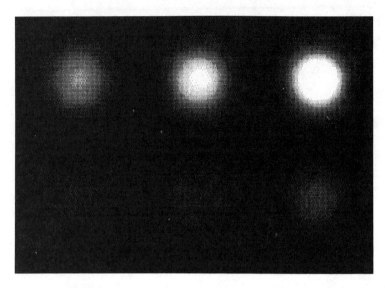

Fig. 9 Box-car image of the flat-bottomed hole sample from which the curves of Figs. 7 and 8 were taken. The first gate was set at 0.75s.

Fig. 10 a) Box-car image of impact damage test sample. b) Box-car image of impact damage in a graphite-epoxy in a second graphite-epoxy test sample.

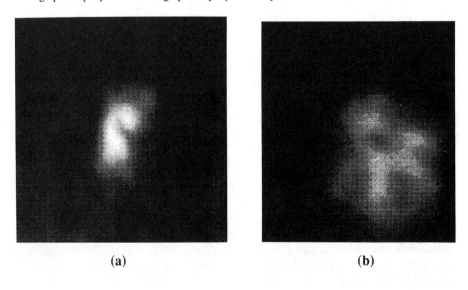

(a) (b)

times, or else use an alternative time-slicing technique which is described later in this chapter under "Thermal Wave Tomography".

The sample shown in Fig. 5 was fabricated to illustrate the behavior of thermal wave echoes from subsurface features. In Figures 10a and 10b we show images of two samples of more practical interest. These consist of graphite-epoxy laminates which had been subjected to impact testing with different loads.

III. PULSED THERMAL-WAVE TOMOGRAPHY

The system described above and depicted in Fig. 4 can be operated in another, quite different, imaging mode. In this mode one makes use of the relationship between the times of arrival of the peaks of the thermal wave echo pulses and the depths of the defects to produce a thermal-wave tomogram of the sample [34,42,43,45]. The word "tomogram" here is used in its original sense, with the meaning of "slicing", not as post-processing computer-assisted reconstruction, which has recently become fashionable [47-49]. The purpose of the present technique is to carry out real-time slicing. Various forms of thermal wave tomography have been carried out by others. [24,47-49] but in all cases they require post-processing, and hence are not real-time techniques. These tomograms are obtained by programming the real-time processor to perform the following sequence of operations. First, a reference area is selected on a featureless area of the sample. This area will be used in real time to produce a dynamic temperature-time reference curve similar to the bottom curve in Fig. 6. Then, immediately following the firing of the flash lamps, the processor begins to acquire curves such as are shown in Fig. 7 by subtracting this dynamic reference from each pixel of the image as it comes into the processor, and storing the resulting echo image. The processor then follows the time evolution of a locally smoothed echo curve for each pixel of the image in order to locate the peak of the echo pulse for the pixel. Once located, this peak time is stored in a frame buffer to produce a peak time image. At the same time, a second buffer is used to store the value of the peak amplitude of the echo, thus producing a "maximum contrast" image which displays each pixel as it appeared at the time when its individual contrast with the background was greatest. The system currently has the ability to perform this sequence of operations with a 10 Hz sampling rate. Since the echo's peak time is related to the depth of the defect, the peak-time image can be "sliced" to produce tomographic displays of the interior of the sample as a series of stacked layers. The maximum contrast image on the other hand, displays features which occurred at different times in the actual time evolution of the image, all in one combined image, and each with the best possible contrast. Figure 11 shows a three-dimensional perspective plot of a tomographic image of the flat-bottomed hole sample shown in Fig. 5. In this image lighter gray-scale and greater height both

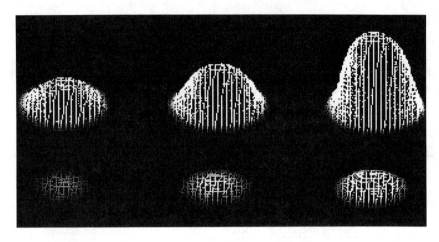

Fig. 11 Perspective plot of a thermal wave tomogram of the flat-bottomed hole sample shown in Fig. 5. Lighter gray-scale and greater height indicate greater depth beneath the heated surface.

indicate greater depth beneath the heated surface. The maximum contrast image corresponding to Fig. 11 is shown in Fig. 12, where lighter gray-scale and greater height now indicate greater peak amplitude of the echo. Notice that where the echo amplitude is greatest, i.e. over the shallowest hole, the peak time is least, and where the peak time is greatest, i.e. over the deepest hole, the echo amplitude is least. This is consistent with Eqs. 7 and 8, and with the time plots of thermal wave echoes shown in Fig. 7. The ring-shaped patterns which appear in the tomographic images of the shallower holes result from three dimensional heat flow and from multiple scattering effects, neither of which is included in the simple theory presented above.

It is instructive to compare the results of thermal wave pulse-echo imaging, with conventional ultrasonic pulse-echo imaging. Such a comparison is shown in Figs. 13a and 13b. Figure 13a shows a thermal wave tomogram of another graphite-epoxy impact damage sample. This particular impact caused some damage directly under the impact site, with concentric rings of damage around that damage. The first ring of damage is shallower than the damage directly under the impact site, with the remaining rings getting progressively deeper in a more-or-less conical pattern. The deepest ring is decidedly asymmetrical. This thermal-wave tomogram can be compared to the ultrasonic pulse-echo image shown in Fig. 13b. This image was obtained by scanning a printed color image of the ultrasonic image, followed by a modification of the resulting gray-scale to obtain the closest correspondence to the depth scale used in the thermal-wave tomogram of Fig. 13a. This particular ultrasonic image happens to have lower resolution than the thermal wave image, but the general features of the two are clearly the same.

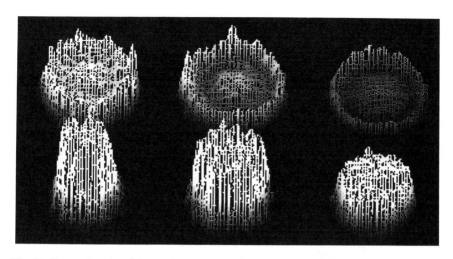

Fig. 12 Perspective plot of the maximum contrast image corresponding to the tomogram shown in Fig. 11. Lighter gray-scale and greater height indicate greater peak amplitude of the thermal wave echo.

Fig. 13 a) Thermal wave tomogram of a graphite-epoxy impact damage specimen. The specimen shows damage directly under the impact site with rings of shallower damage around the central damage, and rings of deeper damage surrounding the shallower damage. b) A pulse-echo ultrasonic image of the damage shown in (a). Although this image happens to have lower resolution than the thermal wave tomogram, the same general features are seen in this image as are seen in the tomogram in (a).

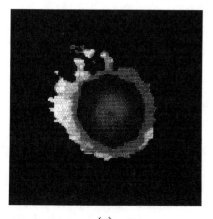
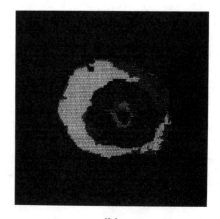

(a)   (b)

IV. INVERSE SCATTERING AND OBJECT RECONSTRUCTION

One criticism that has been directed at thermal wave imaging is that the resolution of the images is not as good as that of high-quality ultrasonic images. The origin of this criticism is the blurring of thermal wave images which occurs as a result of the lateral diffusion of heat while the waves are propagating back to the surface. Ultrasonic images are much less susceptible to such problems because, for many purposes, ultrasound can be regarded as propagating in straight lines. Crowther, et al. [40-42,46] have recently demonstrated a technique for removing the blurring of thermal wave images by means of a simple numerical algorithm based on inverse scattering theory. The algorithm at present is applicable only to planar scatterers, but this covers a very important class of defects of practical interest, namely delaminations and disbonds in laminated composites and adhesively bonded structures.

As is the case with any inverse scattering calculation, an absolute prerequisite for inversion of thermal wave images is an accurate direct scattering model. The inversions described here were obtained through the use of the Born approximation, with appropriate modifications to take account of the strong thermal wave scattering which occurs at delaminations and disbonds. A detailed derivation of this model will be published elsewhere. Its essential feature is that the theoretical expression for a thermal wave echo signal (such as the curves of Fig. 7) in an image can be written as a convolution of two, two-dimensional functions. The first of these describes the propagation of echo back to the surface of the sample. Although it describes more than just the propagation of the pulse from the scatterer to the detector, it serves as a mathematical analog of the "point-spread function" used in optics and ultrasonics. The second function in the convolution is just the function which describes the two-dimensional shape of the scatterer. The result is an equation of the form,

$$\Delta T(x,y,t) = \iint dx'dy' \; g(x-x',y-y',t) \, f(x',y') \; . \qquad (11)$$

Here $\Delta T(x,y,t)$ represents the echo (or equivalently, contrast) signals, comprising the image as a function of time; $g(x,y,t)$ is the "heat-spread function"; and $f(x,y)$ represents the shape of the scatterer. The exact form of $g(x,y,t)$ depends on the approximations used in the model. The inverse scattering calculation is begun by performing a two-dimensional spatial Fast Fourier Transform (FFT) on the experimental echo image, represented in Eq. (11) by $\Delta T(x,y,t)$. The same procedure is then applied to the appropriate expression for $g(x,y,t)$ at a value of t corresponding to the time delay at which the image was obtained. In principle, then, one simply has to divide the first of these FFT's by the second, and do an inverse FFT on the result to obtain the desired shape function $f(x,y)$. However, the experimental image represented by ΔT is not arbitrarily accurate, because it contains both digital and thermal noise. The effect of this noise is to obscure the shape of the scatterer in the image of the inverted FFT. Fortunately, both kinds of noise tend to appear only in relatively high spatial frequencies, and these frequencies can be removed by applying any of a variety of low-pass filters to the FFT of the experimental data. When this is done, the result of the division and the inverse transformation is an image of the shape function $f(x,y)$, with only a slight rounding of the corners due to the loss of high-frequency information. The complete numerical calculation takes about one minute on a Macintosh IIfx. It would take considerably less time if the FFT of the heat-spread function $g(x,y,t)$ were calculated and stored prior to the processing of the experimental image.

In Fig. 14a we show an image of a sample which was constructed to demonstrate the ability of the inversion process to improve the resolution of thermal wave images. The target is again a set of flat bottomed holes in a plastic sample. However, in this case, the holes were placed very close together. Each hole is 12.5 mm in diameter and their centers are 13 mm apart, thus leaving a 1 mm thick wall between them. As it turned out, the heating was not very uniform when the flash lamps were discharged, so there

Fig. 14 a) A pulse-echo thermal wave image of two 12.5 mm diameter flat bottomed holes in a plastic sample. The centers of the holes are 13 mm apart. The asymmetry of the image is due to a gradient in the heating pulse. b) The image produced by applying an inverse scattering algorithm to the image in (a). Note the removal of the asymmetry and the clear resolution of the 1 mm separation of the two holes.

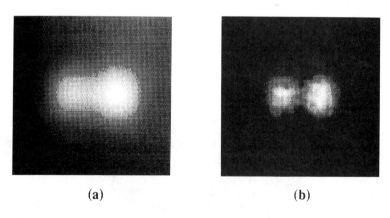

(a)   (b)

was a rather large temperature gradient across the sample. The asymmetry in the size of the two lobes of the image in Fig. 14a is a result of this gradient. In spite of this asymmetry, the inverted image which is shown in Fig. 14b, displays the two holes as very nearly the same size, and with a clearly resolved wall between them.

A second example of the inversion process is shown in Figs. 15 and 16. In this case the sample is a panel of an aircraft aluminum alloy with three equal-size rectangular milled shapes at different depths from the front surface. Figure 15 shows the raw pulse-echo thermal wave image. One should notice the decreasing signal level, decreasing apparent size, and increasingly rounded shapes of these features as their depth increases. The image which results from the application of the inverse scattering algorithm is shown in Fig. 16 using a display method which makes the defects appear to be raised from the surface of the image. Note the clear rectangular shape of the image of each defect and the restoration of their correct (equal) sizes.

V. LOCK-IN VIDEO THERMAL WAVE IMAGING

The system used to effect box-car integration and tomography in pulse-echo video thermal wave imaging can also be used, with slight modification, to carry out lock-in video thermal wave imaging. In this mode, the heat source is periodically modulated in intensity. Figure 17 shows a block diagram of this mode of operation. The system uses the same video camera and real-time processor, but the heat source is now an amplitude modulated laser, an a.c. current in the sample, or some other source of periodic heating. The frequency of the heat source may either be determined

Fig. 15 A thermal wave pulse echo image of a panel of an aircraft aluminum alloy with three equal-size rectangular shapes milled at different depths into the rear surface. The images are seen to become fainter as the depth increases.

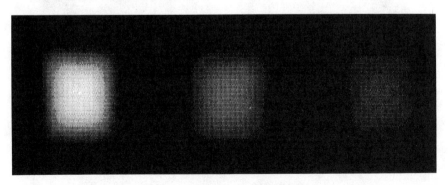

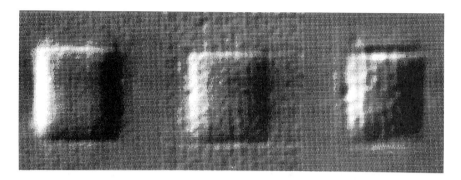

Fig. 16 The result of applying the inverse scattering algorithm to the image of Fig. 15. The image is displayed with a technique which makes the images of the holes appear to be raised from the surface. Note the recovery of the rectangular shape and size of each of the defects.

externally, say by a signal generator, or may be generated internally by deriving it from the synchronization signals of the camera or from the control signals of the digitizer. In the case of externally generated modulation, the heating signals are asynchronous with the camera frame and line frequencies. When the heating frequency is internally generated, the heating and camera frequencies are related to, but not necessarily exactly the same as, at least one of the camera's frequencies. However, in either case, the modulation frequency is fed to the real-time processor as a reference, so that the data processing is *always* synchronous with the heating signal. The signal processing is accomplished by simultaneously multiplying the incoming data stream from the camera by both the sine and cosine of the reference frequency in real time, and averaging each in its own frame buffer. The result is an in-phase lock-in image in one buffer and a quadrature image in the second buffer. These images may then be combined to obtain an image at any given phase, or to obtain separate phase and amplitude images, etc. In these lock-in images, any constant background image or non-synchronous time-dependent image, including image noise, is cancelled. When the modulation frequency is not synchronous with the camera, this cancellation is statistical, i.e. the ratio of the synchronous image signal to the background signal increases like the square root of the number of averages. When the modulation frequency is derived from the camera, the frequency is ordinarily selected such that the background cancellation is exact. The noise cancellation is, of course, necessarily still statistical in nature.

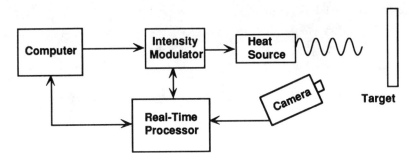

Fig. 17 Block diagram of the Lock-in Video Imaging System at Wayne State University.

In Fig. 18 we show an unsynchronized image of an operating power transistor. In Fig. 19 we show a lock-in thermal wave image of the same operating power transistor, taken at a frequency of 656 Hz. It will be noted that the outside rim of the transistor, whose temperature has only a d.c. component, has disappeared in the lock-in image (compare Figs. 18 and 19). Note also the localization of the a.c. heating in several regions of the device, possibly indicating defects in the structure.

VI. THE FLYING SPOT CAMERA

Figure 20 shows a schematic drawing of the experimental set-up of the flying spot camera. An argon ion laser heating beam passes through a focusing lens, and is deflected by a fixed mirror onto a pair of scanning mirrors. The scanning mirrors are arranged in such a way that they can scan a two-dimensional raster on the sample. A portion of the infrared radiation emitted by the surface of the sample passes back through the scanning mirrors to a mirror which reflects the visible light and is transparent to the infrared radiation. The transmitted IR radiation passes through the fixed mirror and is focused by a germanium lens onto a cooled HgCdTe IR detector. Since the IR radiation had passed through the same scanning mirrors as the laser beam, the point which is imaged on the detector by the germanium lens is scanned in the same pattern as the laser spot, but with a relative displacement which is determined by the position of the detector in the focal plane of that lens. The detector and sample positions are determined by computer-controlled stepping motors. The same computer (a Macintosh II) controls the scanning

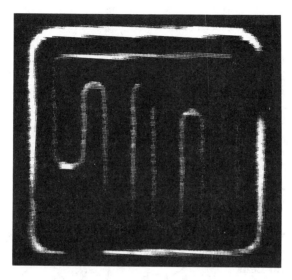

Fig. 18 An unsychronized image of an operating power transistor.

Fig. 19 A lock-in thermal wave image of the same operating power transistor as shown in Fig. 18, taken at a frequency of 656 Hz.

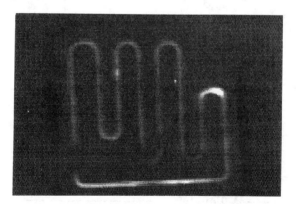

mirrors through a dedicated single board computer and processes the data from the IR detector. One way to think of the operation of the system is to imagine the mirrors as scanning two spots on the surface of the sample, the actual focal spot of the laser beam, which supplies the heat, and the image of the detector, which can be thought of as detecting the thermal waves generated by the heating spot. Thus, the relative position of the two spots, while fixed during any given scan, can be varied in either or both of the two scan directions to measure thermal wave propagation in an arbitrary direction with an arbitrary time delay.

Fig. 20 Block diagram of the Flying Spot Camera.

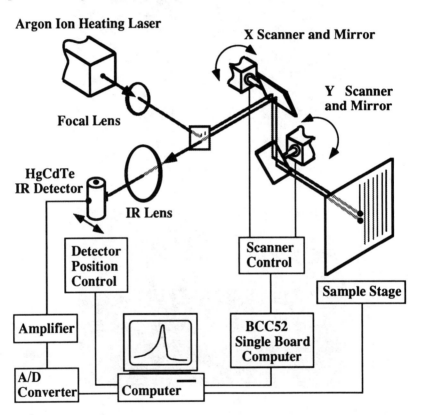

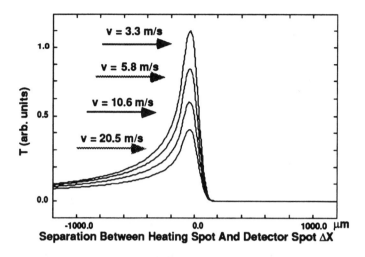

Fig. 21 Temperature profiles of the moving heating spot as a function of the separation, Δx, between the heating spot and the detector spot for a material with a thermal diffusivity, $\alpha = 0.055$ cm^2/s, and for spot speeds varying from 3.3 m/s to 20.5 m/s. The assumed source and detector sizes are $w_s = 100$ μm and $w_d = 100$ μm, respectively. The corresponding experimental curves are shown in Fig. 22.

Fig. 22 Experimental temperature profiles for a graphite specimen with $\alpha = 0.055$ cm^2/s, and for spot speeds varying from 3.3 m/s to 20.5 m/s.

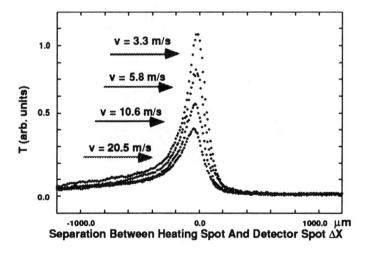

The measured temperature profile, $\tilde{T}$, of the moving laser spot is the convolution of temperature distribution, $T(r,t)$, given by Eq. (10), with a Gaussian source distribution of width w_s, and with a detector response function, D,

$$\tilde{T}(\upsilon, w_s, w_d) = T(r,t) \otimes G(r,t,w_s) \otimes D(r,t,w_d). \quad (12)$$

Here, υ represents the source speed, and w_d represents the width of the (square) detector.

Using Equation (12), we have calculated the temperature profiles of the moving heating spot as a function of the separation, Δx, between the heating spot and the detector spot for a material with a thermal diffusivity $\alpha = 0.055$ cm^2/s, and for spot speeds varying from 3.3 m/s to 20.5 m/s. The assumed source and detector sizes are $w_s = 100$ μm and $w_d = 100$ μm, respectively. The results of our calculation are shown in Fig. 21.

As example of an image created with the flying spot camera, in Fig. 23 we show an image of a graphite sample containing a circular saw cut which extends through the top of the sample. Just below the opening of the

Fig. 23 Flying spot thermal wave image of a fractured region in the vicinity of a circular saw cut in a graphite sample.

cut one can see a strong thermal wave reflection from a subsurface fracture region, together with a very narrow crack extending from the left side of that region, down towards the bottom of the image.

VII. REFERENCES

1. For an earlier review of thermal wave imaging for non-destructive evaluation, see for example, R.L. Thomas, L.D. Favro and P.K. Kuo, Can. J. Phys., **64**, 1234 (1986).

2. Y.Q. Wang, P.K. Kuo, L.D. Favro and R.L. Thomas, *Photoacoustic and Photothermal Phenomena II,* (J.C. Murphy, J.W. Maclachlan Spicer, L.C. Aamodt, and B.S.H. Royce, Eds.) 24 (Springer-Verlag, Berlin, 1990).

3. W.N. Reynolds, Can. J. Phys., **64**, 1150 (1986).

4. P. Cielo, X. Maldague, A.A. Deom and R. Lewak, Mater. Eval., **45**, 452 (1987).

5. J. Jaarinen, C.B. Reyes, I.C. Oppenheim, L.D. Favro, P.K. Kuo and R.L. Thomas, *Rev. Progr. Quant. NDE*, (D.O. Thompson and D. Chimenti Eds.) 1347 (Plenum, New York, 1987).

6. J.C. Murphy, J.W. Maclachlan and L.C. Aamodt, *Rev. Progr. Quant. NDE*, (D.O. Thompson and D. Chimenti Eds.) 245 (Plenum, New York, 1988).

7. D. Balageas, A. Deom and D. Boscher, *Eurotherm* **4**, Nancy, June 28-July 1, (1988).

8. P.K. Kuo, J. Hartikainen, I.C. Oppenheim, L.D. Favro, Z.J. Feng and R.L. Thomas, *Rev. Progr. Quant. NDE*, (D.O. Thompson and D. Chimenti, Eds.) 273 (Plenum, New York, 1988).

9. P.K. Kuo, I.C. Oppenheim, L.D. Favro, Z.J. Feng, R.L. Thomas, J. Hartikainen and L.J. Inglehart, *Photoacoustic and Photothermal Phenomena*, (P. Hess and J. Pelzl Eds.) 496 (Springer-Verlag, Berlin, 1988).

10. J. Jaarinen, J. Hartikainen and M. Luukkala, *Rev. Progr. Quant. NDE*, (D.O. Thompson and D. Chimenti, Eds.) 1311 (Plenum, New York, 1989).

11. J.W. Maclachlan, L.C. Aamodt and J.C. Murphy, *Rev. Progr. Quant. NDE*, (D.O. Thompson and D. Chimenti, Eds.) 1297 (Plenum, New York, 1989).

12. J.S. Heyman, W.P. Winfree, F.R. Parker, D.M. Heath and C.S. Welch, *Non-destructive Monitoring of Materials Properties Symposium*, (J. Holbrook and J. Bussiere, Eds.) 211 (1989).

13. P.K. Kuo, T. Ahmed, L.D. Favro, H.J. Jin and R.L. Thomas, *Proc. 17th Symposium Nondestr. Eval.*, 238 (Southwest Research Institute, San Antonio, 1989).

14. P.K. Kuo, T. Ahmed, L.D. Favro, H-J. Jin, R.L. Thomas, J. Jaarinen and J. Hartikainen, *Rev. Progr. Quant. NDE*, (D.O. Thompson and D. Chimenti, Eds.) 1305 (Plenum, New York, 1989).

15. T. Ahmed, P.K. Kuo, L.D. Favro, H-J Jin, R.L. Thomas and R. Dickie, *Rev. Progr. Quant. NDE*, (D.O. Thompson and D. Chimenti, Eds.) 1385 (Plenum, New York, 1989).

16. P.K. Kuo, T. Ahmed, L.D. Favro, H-J Jin and R.L. Thomas, J. Nondestr. Eval., **8**, 97 (1989).

17. J.D. Morris, D.P. Almond, P.M. Patel and H. Reiter, *Photoacoustic and Photothermal Phenomena II*, (J.C. Murphy, J.W. Maclachlan-Spicer, L.C. Aamodt and B.S.H. Royce, Eds.) 71 (Springer-Verlag Berlin, 1990).

18. S.K. Lau, D.P. Almond and P.M. Patel, *Photoacoustic and Photothermal Phenomena II*, (J.C. Murphy, J.W. Maclachlan-Spicer, L.C. Aamodt and B.S.H. Royce, Eds.) 74 (Springer-Verlag Berlin, 1990).

19. X. Maldague, P. Cielo, D. Poussart, D. Craig and R. Bourret, *Proc. SPIE Thermosense XII*, **1313**, 161 (1990).

20. M. Connolly and D. Copley, Mater. Eval., **48**, 1461 (1990).

21. J. Hartikainen and M. Luukkala, *Photoacoustic and Photothermal Phenomena II*, (J.C. Murphy, J.W. Maclachlan-Spicer, L.C. Aamodt and B.S.H. Royce), 496 (Springer-Verlag Berlin, 1990).

22. T. Ahmed, P.K. Kuo, L.D. Favro, H-J. Jin, P. Chen and R.L. Thomas, *Rev. Progr. Quant. NDE*, (D.O. Thompson and D. Chimenti, Eds.)

2001 (Plenum, New York, 1990).

23. L.C. Aamodt, J.W. Maclachlan Spicer and J.C. Murphy, *Photoacoustic and Photothermal Phenomena II*, (J.C. Murphy, J.W. Machlachlan-Spicer, L. Aamodt and B.S.H. Royce, Eds.) 59 (Springer-Verlag Berlin, 1990).

24. V. Vavilov, T. Ahmed, H.J. Jin, R.L. Thomas and L.D. Favro, Sov. J. NDT, **12**, 60 (1990).

25. L.D. Favro, T. Ahmed, H.J. Jin , P.K. Kuo and R.L. Thomas, *Proc. SPIE Thermosense XII*, **1313**, 302 (1990).

26. L.D. Favro, T. Ahmed, H.J. Jin, P. Chen, P.K. Kuo, and R.L. Thomas, in *Photoacoustic and Photothermal Phenomena II*, (J.C. Murphy, J.W. Maclachlan-Spicer, L.C. Aamodt and B.S.H. Royce, Eds.) 490 (Springer-Verlag Berlin, 1990).

27. T. Ahmed, H.J. Jin, P. Chen, P.K. Kuo, L.D. Favro and R.L. Thomas, *Photoacoustic and Photothermal Phenomena II*, (J.C. Murphy, J.W. Maclachlan-Spicer, L.C. Aamodt and B.S.H. Royce, Eds.) 30 (Springer-Verlag Berlin, 1990).

28. D. Burleigh and W. De La Torre, *Proc. SPIE Thermosense XIII*, **1467**, 303 (1991).

29. C. Hobbs, D. Kenway-Jackson and J. Milne, *Proc. SPIE Thermosense XIII*, **1467**, 264 (1991).

30. J. Rantala and J. Hartikainen, *Rev. Prog. Quant. Nondestr. Eval.* **10A**, 1051 (1991).

31. J.W. MacLachlan Spicer, W.D. Kerns, L.C. Aamodt and J.C. Murphy, *Rev. Prog. Quant. NDE,* **10B**, 1193 (1991).

32. L.D. Favro, H.J. Jin, T. Ahmed, X. Wang, P.K. Kuo and R.L. Thomas, in *Rev. Progr. Quant. NDE*, (D.O. Thompson and D. Chimenti, Eds.) **10B**, 1201 (Plenum, New York, 1991).

33. L.D. Favro, T. Ahmed, D. Crowther, H.J. Jin, P.K. Kuo, R.L. Thomas and X. Wang, *Proc. SPIE Thermosense XIII*, **1467**, 290; and 132 (1991).

34. L.D. Favro, H.J. Jin, Y.X. Wang, T. Ahmed, X. Wang, P.K. Kuo and

R.L. Thomas, *Rev. Progr. Quant. NDE*, (D.O. Thompson and D. Chimenti, Eds.) **11** (Plenum, New York, to be published).

35. P.K. Kuo, L.D. Favro and R.L. Thomas, *Proc. 1991 SEM Spring Conf. on Exp. Mechanics*, 244 (1991).

36. S.K. Lau, D. Almond and P.M. Patel, *Digest, 7th Int. Topical Meet. Photoacoustic Photothermal Phenomena*, (Doorwerth, The Netherlands, 1991).

37. R. Lehtiniemi, J. Hartikainen, J. Rantala, J. Varis and M. Luukkala, *Proc. 18th Annual Rev. Progr. Quant. Nondestr. Eval.*, (D.O. Thompson and D. Chimenti, Eds.) **11** (Plenum, New York, to be published).

38. R. Lehtiniemi, J. Hartikainen, J. Varis and M. Luukkala, *Digest, 7th Int. Topical Meet. Photoacoustic Photothermal Phenomena*, (Doorwerth, The Netherlands, 1991).

39. J.W.M. Spicer, L.C. Aamodt, W.D. Kerns and J.C. Murphy, *Digest, 7th Int. Topical Meet. Photoacoustic Photothermal Phenomena*, (Doorwerth, The Netherlands, 1991).

40. D. Crowther, L.D. Favro and R.L. Thomas, *Proc. 18th Annual Rev. Progr. Quant. Nondestr. Eval.*, (D.O. Thompson and D. Chimenti, Eds.) **11** (Plenum, New York, to be published).

41. D.J. Crowther, L.D. Favro, P.K. Kuo and R.L. Thomas, Digest; *7th Int. Topical Meet. Photoacoustic Photothermal Phenomena*, (Doorwerth, The Netherlands, 1991).

42. L.D. Favro, H.J. Jin, Y.X. Wang, T. Ahmed, X. Wang, P.K. Kuo and R.L. Thomas, *Proc. 18th Annual Rev. Progr. Quant. Nondestr. Eval.*, (D.O. Thompson and D. Chimenti, Eds.) **11** (Plenum, New York, to be published).

43. L.D. Favro, H.J. Jin, P.K. Kuo, R.L. Thomas and Y.X. Wang, *Digest, 7th Int. Topical Meet. Photoacoustic Photothermal Phenomena* (Doorwerth, The Netherlands, 1991).

44. L.D. Favro, D.J. Crowther, P.K. Kuo and R.L. Thomas, *Proc. SPIE Thermosense XIV*, **1682**, 178 (1992).

45. R.L. Thomas, L.D. Favro, H.J. Jin, P.K. Kuo, and Y.X. Wang, *Proc.*

14th Int. Congress Acoustics, **1**, A13-8 (Beijing, China, 1992).

46. L.D. Favro, D.J. Crowther, P.K. Kuo and R.L. Thomas, *Proc. 14th Int. Congress Acoustics*, **1**, A13-5 (Beijing, China, 1992).

47. M. Munidasa and A. Mandelis, J. Opt. Soc. Am. **A8**, 1851 (1991).

48. A. Mandelis, J. Phys. A: Math. Gen. **24**, 2485 (13991).

49. M. Munidasa, A. Mandelis and C. Ferguson, Appl. Phys. **A 54**, 244 (1992).

THEORY OF THERMAL WAVE SURFACE TEMPERATURES AND SURFACE DISPLACEMENTS FOR BULK AND LAYERED SAMPLES IN ONE AND THREE DIMENSIONS

Jon Opsal and Allan Rosencwaig

Therma-Wave, Inc.
Fremont, CA 94539

I.	INTRODUCTION	54
II.	THEORY	54
	1. Temperature in Bulk Sample	55
	2. Thermoelastic Displacement and Deflection in a Bulk Sample	59
	3. Temperature in a Layered Sample	64
	4. Thermoelastic Displacement and Deflection in a Layered Sample	65
III.	DISCUSSION	68
IV.	REFERENCES	69

I. INTRODUCTION

Thermal wave techniques are widely used for studying a wide variety of materials and samples [1-7]. One of the most important applications of thermal wave physics is the study of samples with a surface layer or film. The most common thermal wave techniques for the study of such samples are those that are dependent on the sample surface temperature (optical beam deflection [8,9], photothermal radiometry [10,11], modulated reflectance [12,13], or in the thermoelastic expansion of the sample surface (modulated surface displacement [14,15] and modulated surface deflection [16,17], which measures the local slope of the thermoelastic deformation).

There have been a number of theoretical treatments of the thermal wave process [17-24]. However it is difficult to extract from these theories those physical insights that are useful to the design and interpretation of experiments. In this analysis we present a rigorous, unified theory for both surface temperatures and surface thermoelastic expansion of bulk and layered samples in both one and three dimensions. We are particularly interested in the thin single layer sample wherein the layer differs from the underlying substrate principally in its thermal conductivity. This is a particularly important class of samples because it closely approximates the situation where the surface of a material, such as a semiconductor, is altered by some physicochemical process such as ion implantation, etching, etc. Our goal is to gain some basic physical insight into how the different thermal wave signals change in going from bulk to layered samples, how they change in going from the one to the three dimensional regime, and how they compare in sensitivity to the presence of the layer.

II. THEORY

In this section we will develop the theory for modulated reflectance and modulated thermoelastic expansions in both one and three dimensions. The one-dimensional regime is when the spot size is much larger than the thermal diffusion length and the three-dimensional regime is when the spot size is much smaller than the thermal diffusion length. We will deal with both bulk samples and samples with a thin surface layer. We will assume that this layer differs from the underlying substrate only in its thermal conductivity and that the layer thickness is always small compared to either the heating spot size or any of the thermal diffusion lengths. In this analysis we will also assume that all of the energy from the heating beam is absorbed at the top surface of the sample. Figure 1 schematically depicts the model structure that we will be using. The symbols used in this Figure are defined in the analysis that follows.

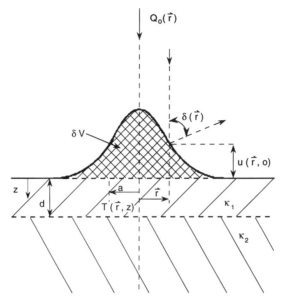

Fig. 1 Schematic depiction of model structure used in the theory. The thickness d of the upper layer is actually assumed to be < a in the theory.

1. Temperature in Bulk Sample

We first need to find the temperature $T(r,z)$ as a function of lateral position $r = (x,y)$ and depth z within the material (Fig. 1). This means solving the thermal wave equation [12-15]:

$$\nabla^2 T + q^2 T = 0 \tag{1}$$

subject to the boundary condition:

$$-\kappa \nabla T \big|_{z=0} = Q_0 f(r) \tag{2}$$

where $Q_0 f(r)$ is the heat flux absorbed at the front surface $z = 0$, the spatial dependence of which is described by $f(r)$. In Equation (1), q is the thermal propagation wavenumber defined by:

$$q = (1 + i)\sqrt{\frac{\omega \rho C}{2\kappa}} \equiv (1 + i)/\mu \tag{3}$$

where ω is the angular modulation frequency, ρ is the density, C is the specific heat, κ is the thermal conductivity, and $\mu = \sqrt{\dfrac{2\kappa}{\omega\rho C}}$ is the thermal diffusion length.

In order to solve this three-dimensional problem we first take advantage of the planar symmetry of layered structures and express $T(r,z)$ and $f(r)$ in terms of their respective Fourier transforms $\tilde{T}(q_r,z)$ and $\tilde{f}(q_r)$. That is,

$$T(r,z) = \int d^2q_r e^{iq_r \cdot r} \tilde{T}(q_r,z) \qquad (4)$$

and

$$f(r) = \int d^2q_r e^{iq_r \cdot r} \tilde{f}(q_r) \qquad (5)$$

From these expressions and Eqs. (1,2) we obtain the one-dimensional differential equation for $\tilde{T}(q_r,z)$

$$\frac{d^2\tilde{T}}{dz^2} + q_z^2 \tilde{T} = 0 \qquad (6)$$

with the boundary condition

$$-\kappa \frac{d\tilde{T}}{dz} = Q_0 \qquad (7)$$

and where

$$q_z = \sqrt{q^2 - q_r^2} \qquad (8)$$

is the z-component of the thermal wave vector.

From Equations (6,7) we then obtain

$$\tilde{T}(q_r,z) = \frac{iQ_0 \tilde{f}(q_r)}{\kappa q_z} e^{iq_z z} \qquad (9)$$

Substituting Equation (9) into Eq. (4) we obtain

$$T(r,z) = \frac{iQ_0}{\kappa} \int d^2q_r e^{iq_r \cdot r} \bar{f}(q_r) \frac{e^{iq_z z}}{q_z}, \tag{10}$$

a result in a form particularly convenient for understanding and analyzing thermal waves in layered structures. Physically, Equation (10) states that a thermal wave in three-dimensions can be considered as a linear superposition of plane waves, the propagation of which in layered media is well understood. We should also point out that no assumptions regarding the symmetry of the source term $f(r)$ have yet been made. If we now assume circular symmetry, as often occurs in practice, then Eq. (10) reduces to the Hankel transform

$$T(r,z) = \frac{2\pi i Q_0}{\kappa q} \int q_r dq_r J_0(q_r r) \bar{f}(q_r) \frac{e^{iq_z z}}{q_z} \tag{11}$$

where J_0 is the Bessel function of order zero.

As a way of understanding these results we consider the behavior of $T(r,z)$ in the two extremes of high and low frequency. Letting a denote the effective radius of the source term, then in the high frequency or one-dimensional limit, $a \gg \mu$ and consequently throughout the range of integration for which $\tilde{f}(q_r)$ is significant in Eq. (10), $q_z \approx q$. Therefore

$$T(r,z) = \frac{iQ_0 f(r)}{\kappa q} e^{iqz} \tag{12}$$

and at the surface

$$T(r,0) = \frac{iQ_0 f(r)}{\kappa q}$$
$$= \frac{(1+i)Q_0 f(r)}{\sqrt{2\kappa \omega \rho C}} \tag{13}$$

Thus in the one-dimensional limit the temperature distribution at the surface $z = 0$ simply maps the heat source distribution $f(r)$ and varies as $\kappa^{-1/2}$ and $\omega^{-1/2}$.

In order to understand the temperature distribution in the three-dimensional case we evaluate the integral in Eq. (11) in the low frequency or three-dimensional limit $a \ll \mu$ and thus in the limit $q_z \approx iq_r$. Assuming a Gaussian profile in the source term as is often the case

$$f(r) = e^{-r^2/a^2} \tag{14}$$

and

$$\tilde{f}(q_r) = \frac{a^2}{4\pi} e^{-\frac{q_r^2 a^2}{4}} \tag{15}$$

Substituting Equation (14) into (11) we then obtain in this low frequency or 3-dimensional limit

$$T(r,z) = \frac{Q_0 a}{2\kappa} \int dq_r J_0(q_r r) e^{-\frac{q_r^2 a^2}{4}} e^{-q_r z} \tag{16}$$

Note that Equation (16) is independent of ω. This is true for the extreme 3-dimensional case since $a \ll \mu$ is equivalent to $\omega \to 0$.

At the surface $z = 0$, the temperature distribution in the low frequency three-dimensional limit is given by

$$T(r,z) = \frac{\sqrt{\pi} Q_0 a}{2\kappa} I_0\left(\frac{r^2}{2a^2}\right) e^{-r^2/2a^2} \tag{17}$$

where I_0 is the modified Bessel function of order zero, $I_0(x) = J_0(ix)$.

At the center of the beam ($r = 0$),

$$T(0,0) = \frac{\sqrt{\pi} Q_0 a}{2\kappa} \tag{18}$$

while for $r \gg a$ and $\mu \gg r$,

$$T(r,0) = \frac{\sqrt{\pi} Q_0 a}{2\kappa} \left(\frac{1}{\sqrt{\pi}} \frac{a}{r}\right)$$
$$= \frac{Q_0 a^2}{2\kappa r} \tag{19}$$

Note that the 3-dimensional temperature drops off slowly as $1/r$ and is still finite for $r \gg a$. This is unrealistic. For finite samples and when heat losses due to radiation, conduction and convection to the surroundings are included, the temperature will drop off faster than $1/r$ in the region $r \gg a$.

2. Thermoelastic Displacement and Deflection in a Bulk Sample

For the thermoelastic deformation of the surface we consider the displacement, $u(r,z)$, which can be derived from a potential function, $\phi(r,z)$

$$u = -\nabla \phi \tag{20}$$

where ϕ is a solution of the thermoelastic wave equation

$$\nabla^2 \phi + q_l^2 \phi = \gamma T. \tag{21}$$

q_l is the longitudinal sound propagation wavenumber

$$q_l = \frac{\omega}{c_l}, \tag{22}$$

where c_l is the longitudinal sound velocity. γ is the thermoelastic coupling constant

$$\gamma = \left(\frac{1+\nu}{1-\nu}\right)\alpha \tag{23}$$

where α is the linear thermal expansion coefficient and ν is Poisson's ratio. In this analysis we are neglecting the presence of shear waves, but as shown in Ref. [24] shear waves add only a small correction to the displacement in the vicinity of the thermal waves.

As in the thermal wave problem we express $\phi(r,z)$ and $u(r,z)$ in terms of their Fourier transforms $\tilde{\phi}(q_r,z)$ and $\tilde{u}(q_r,z)$

$$\phi(r,z) = \int d^2q_r e^{iq_r \cdot r} \tilde{\phi}(q_r,z) \tag{24a}$$

$$u(r,z) = \int d^2q_r e^{iq_r \cdot r} \tilde{u}(q_r,z). \tag{24b}$$

By use of Equation (21), Eqs. (24a,b) lead to the 1-dimensional wave equation

$$\frac{d^2\tilde{\phi}}{dz^2} + q_{lz}^2 \tilde{\phi} = \gamma \tilde{T}, \tag{25}$$

where q_{lz} is the z-component the longtitudinal sound wave propagation vector

$$q_{lz} = \sqrt{q_l^2 - q_r^2}, \qquad (26)$$

with the condition for a stress-free surface given by

$$\tilde{\phi}(q_r, 0) = 0. \qquad (27)$$

Using Equations (6) and (25) we obtain

$$\tilde{\phi}(q_r, z) = \tilde{\phi}_l(q_r, z) - \frac{\gamma \tilde{T}(q_r, z)}{q_z^2 - q_{lz}^2} \qquad (28)$$

where $\tilde{\phi}_l(q_r, z)$ is a solution of the homogenous wave equation

$$\frac{d^2 \tilde{\phi}_l}{dz^2} + q_{lz}^2 \tilde{\phi}_l = 0 \qquad (29)$$

and from Eq. (27) we then have for the stress-free boundary condition

$$\tilde{\phi}_l(q_r, 0) = \frac{\gamma \tilde{T}(q_r, 0)}{q_z^2 - q_{lz}^2}. \qquad (30)$$

Substituting for $\tilde{T}(q_r, z)$ from Eq. (9) we then obtain

$$\tilde{\phi}(q_r, z) = \frac{-i\gamma Q_0 \tilde{f}(q_r)}{\kappa q_z (q_z^2 - q_{lz}^2)} (e^{iq_z z} - e^{iq_u z}) \qquad (31)$$

which implies the vertical displacement

$$\tilde{u}_z(q_r, z) = \frac{-\gamma Q_0 \tilde{f}(q_r)}{\kappa q_z (q_z^2 - q_{lz}^2)} [q_z e^{iq_z z} - q_{lz} e^{iq_u z}]. \qquad (32)$$

At the surface then

$$\bar{u}_z(\mathbf{q}_r, 0) = \frac{-\gamma Q_0 \bar{f}(\mathbf{q}_r)}{\kappa q_z(q_z + q_{lz})} \tag{33}$$

and using Eq. (24) we obtain the front surface displacement as a function of r:

$$u_z(r,0) = \frac{-\gamma Q_0}{\kappa} \int d^2 q_r e^{i\mathbf{q}_r \cdot \mathbf{r}} \frac{\bar{f}(\mathbf{q}_r)}{q_z(q_z + q_{lz})}. \tag{34}$$

Before we proceed with obtaining the explicit expressions for $u_z(r,0)$ in the 1-dimensional and 3-dimensional regimes, it would be of interest to calculate the integral of $u_z(r,0)$ over the surface, that is, the total displaced volume δV

$$\delta V = \int d^2 r \, u_z(r,0) = \frac{-4\pi^2 \gamma Q_0 \bar{f}(0)}{\kappa q (q + q_l)}. \tag{35}$$

Since at all modulations of practical significance $|q| \gg q_l$,

$$\delta V = \frac{-4\pi^2 \gamma Q_0 \bar{f}(0)}{\kappa q^2} = \frac{i 4\pi^2 \gamma Q_0 \bar{f}(0)}{\omega \rho C}$$

$$= \frac{i \gamma Q_0 \pi a^2}{\omega \rho C}, \tag{36}$$

which is independent of κ but varies as ω^{-1}. Note that this expression for δV is exact for both one and three dimensions. Thus under all circumstances the integrated displaced volume for a bulk material is independent of the thermal conductivity.

This result can also be obtained by performing the volume integral on Eq. (21) directly and applying the limit $q_l \to 0$.

This result is to be expected since the total displaced volume δV is simply proportional to the total integrated thermoelastic strain. The total energy input into the material is given by $\int d^2 r \, Q_0 f(r)$. This input energy is converted into thermal energy through the heat capacity ρC of the material and then into strain through the thermoelastic coupling constant γ. Thus the integrated strain and hence the total displaced volume is dependent only on the total input energy, the heat capacity ρC and the thermoelastic coupling constant γ, *and is independent of the thermal conductivity, a transport property.*

The divergence of δV with decreasing ω is the result of $1/r$ dependence of the 3-dimensional temperature distribution. This divergence will disappear when either the finite size of the sample or the heat loss to the surrounding environment is taken into account.

Let us now derive $u_z(r,0)$ for the one-dimensional and three-dimensional cases.

In the high frequency one-dimensional limit $q_z \approx q$ and $q_{lz} \approx q_l$ and Eq. (34) becomes

$$u_z(r,0) = \frac{-\gamma Q_0}{\kappa q(q + q_l)} f(r) . \tag{37}$$

Since at all modulations of practical significance $|q| >> q_l$,

$$u_z(r,0) = \frac{-\gamma Q_0 f(r)}{\kappa q^2} = \frac{i\gamma Q_0 f(r)}{\omega \rho C} . \tag{38}$$

Thus in the one-dimensional limit the surface displacement at any r is independent of thermal conductivity κ and maps the heating source term $f(r)$. It also varies as ω^{-1}.

Assuming that $f(r) = e^{-r^2/a^2}$, we obtain for the low frequency three-dimensional limit

$$u_z(r,0) = \frac{-\gamma Q_0 a^2}{4\pi \kappa} \int d^2 q_r e^{i q_r \cdot r} \frac{e^{-q_r^2 a^2/4}}{q_z(q_z + q_{lz})}$$

$$= \frac{-\gamma Q_0 a^2}{2\kappa} \int q_r J_0(q_r r) \frac{e^{-q_r^2 a^2/4}}{q_z(q_z + q_{lz})} dq_r . \tag{39}$$

Since this integral has a logarithmic divergence we will calculate $u_z(r,0)$ from the deflection $\delta(r)$.

Let us now calculate the local thermoelastic deflection $\delta(r)$ where $\delta(r)$ is given by

$$\delta(r) = -\frac{\partial}{\partial r} u_z(r,0) \tag{40}$$

In the 1-D limit let us use the 1-D limit for $u_z(r,0)$, Eq. (38), to obtain

$$\delta(r,0) = \frac{\gamma Q_0}{\kappa q^2} f'(r) = \frac{-i\gamma Q_0}{\omega \rho C} f'(r) \qquad (41)$$

where $f'(r) = \frac{\partial f(r)}{\partial r}$. The local thermoelastic slope in the 1-D limit thus simply maps out the slope of the source term $f(r)$ and also varies as ω^{-1}.

For the three-dimensional limit and assuming that $f(r) = e^{-r^2/a^2}$, we then obtain from Eqs. (39) and (40)

$$\delta(r,0) = \frac{-\gamma Q_0 a^2}{\kappa} \int q_r^2 J_1(q_r r) \frac{e^{-\frac{q_r^2 a^2}{4}}}{q_z(q_z + q_{lz})} dq_r \qquad (42)$$

Also in the three-dimensional limit, we can set $q_z \approx iq_r$ and $q_{lz} \approx iq_r$, and then obtain

$$\delta(r,0) = \frac{\gamma Q_0 a^2}{2\kappa} \int dq_r J_1(q_r r) \frac{e^{-q_r^2 a^2}}{r}$$

$$= \frac{\gamma Q_0 a}{2\kappa} \left(\frac{a}{r}\right) [1 - e^{-r^2/a^2}] . \qquad (43)$$

For the displacement we then have

$$u_z(r,0) = \int dr \frac{\partial u_z}{\partial r}$$

$$= \frac{\gamma Q_0 a^2}{4\kappa} \left\{ \frac{1}{2} E_i\left(-\frac{r^2}{a^2}\right) - \ln\left(\frac{r}{a}\right) \right\} \qquad (44)$$

The $\ln\left(\frac{r}{a}\right)$ divergence for very large r is a consequence of not including the finite size of the sample or any thermal loss terms such as radiation. Once these terms are included, this divergence disappears.

For $r < a$

$$u_z(r,0) = \frac{\gamma Q_0 a^2}{4\kappa}\left\{\frac{\gamma_1}{2} - \frac{1}{2}\frac{r^2}{a^2}\right\} \qquad (45)$$

where γ_1 is Euler's constant = 0.57722, and at $r = 0$

$$u_z(0,0) = \frac{\gamma Q_0 a^2 \gamma_1}{8\kappa}. \qquad (46)$$

Thus in the three-dimensional limit u_z is independent of frequency and varies as κ^{-1}.

3. Temperature in a Layered Sample

Now we consider the problem of a single layer on a half-space, Fig. (1).

Let us further assume, as shown in this figure, that the only significant difference between the layer and the substrate is the thermal conductivity. Material damaging processes such as ion implantation will affect first a transport property like thermal conductivity more than a thermodynamic property such as specific heat. For the thermal wave in the layer we then have

$$\bar{T}_1(q_r,z) = A_1(e^{iq_u z} + R_1 e^{-iq_u z}) \qquad (47)$$

where

$$A_1 = \frac{iQ_0 \bar{f}(q_r)}{\kappa_1 q_{1z}(1 - R_1)} \qquad (48)$$

and

$$R_1 = \left(\frac{\kappa_1 q_{1z} - \kappa_2 q_{2z}}{\kappa_1 q_{1z} + \kappa_2 q_{2z}}\right) e^{2iq_u d} \qquad (49)$$

For a thin film $q_{1z} d \ll 1$ we find:

$$\tilde{T}_1(\boldsymbol{q}_r,0) = \frac{iQ_0}{\kappa_2 q_{2z}}\left\{1 + iq_{1z}d\left(\frac{\kappa_1 q_{1z}}{\kappa_2 q_{2z}} - \frac{\kappa_2 q_{2z}}{\kappa_1 q_{1z}}\right)\right\}. \quad (50)$$

In the one-dimensional limit this becomes

$$T_1(r,0) = \frac{iQ_0 f(r)}{\kappa_2 q_2}\left\{1 + iq_1 d\left(\frac{\kappa_1 q_1}{\kappa_2 q_2} - \frac{\kappa_2 q_2}{\kappa_1 q_1}\right)\right\}$$

$$= \frac{(1+i)Q_0 f(r)}{\kappa_2 q_2}\left\{1 + (1-i)\frac{d}{\mu_1}\left(\frac{\sqrt{\kappa_2}}{\sqrt{\kappa_1}} - \frac{\sqrt{\kappa_1}}{\sqrt{\kappa_2}}\right)\right\} \quad (51)$$

In the three-dimensional limit q_{1z} and $q_{2z} \to iq_r$:

$$T_1(r,0) = \frac{Q_0\sqrt{\pi}\,a}{2\kappa_2}\left\{I_0\left(\frac{r^2}{2a^2}\right)e^{-r^2/2a^2}\right.$$

$$\left. + \frac{2}{\sqrt{\pi}}\left(\frac{d}{a}\right)\left(\frac{\kappa_2}{\kappa_1} - \frac{\kappa_1}{\kappa_2}\right)e^{-r^2/a^2}\right\} \quad (52)$$

As in the case of bulk samples the temperature varies as $\omega^{-1/2}$ in the one-dimensional limit and is independent of ω in the extreme three-dimensional limit.

4. Thermoelastic Displacement and Deflection in a Layered Sample

For the thermoelastic potential in the layer we have

$$\tilde{\phi}_1(\boldsymbol{q}_r,z) = B\exp(iq_{1z}z) - \frac{\gamma \tilde{T}_1(\boldsymbol{q}_r,z)}{q_{1z}^2 - q_{lz}^2}, \quad (53)$$

which for the stress-free boundary condition $\tilde{Q}_1(\boldsymbol{q}_r,0) = 0$ implies

$$B = \frac{\gamma \tilde{T}_1(q_r,0)}{q_{1z}^2 - q_{lz}^2}, \qquad (54)$$

and, therefore, that

$$\tilde{\phi}_1(q_r,z) = \frac{\gamma_1 A_1}{q_{1z}^2 - q_{lz}^2}\{(1+R_1)\exp(iq_{lz}z) - [\exp(iq_{lz}z) + R_1\exp(-iq_{lz}z)]\}. \qquad (55)$$

This, in turn, leads to the surface displacement

$$\bar{u}_{1z}(q_r,0) = \frac{-i\gamma A_1}{q_{1z}^2 - q_{lz}^2}[q_{lz}(1+R_1) - q_{1z}(1-R_1)], \qquad (56)$$

which can be rewritten as

$$\bar{u}_{1z}(q_r,0) = i\gamma A_1\left[\frac{1}{q_{1z}+q_{lz}} - \frac{R_1}{q_{1z}-q_{lz}}\right]. \qquad (57)$$

As we did earlier, let us at this point calculate the total displaced volume δV

$$\delta V = \int d^2r\, u_{1z}(r,0)$$

$$= i\gamma \int d^2r\, d^2q_r\, e^{iq_r \cdot r} \tilde{f}(q_r) A_1 \left[\frac{1}{q_{1z}+q_{lz}} - \frac{R_1}{q_{1z}-q_{lz}}\right]$$

$$= 4\pi^2 i\gamma A_1 \tilde{f}(0)\left[\frac{1}{q_1+q_l} - \frac{R_1}{q_1-q_l}\right]. \qquad (58)$$

Given that $q_l \ll |q_1|$, Eq. (58) yields:

$$\delta V = \frac{-4\pi^2 \gamma Q_0 \tilde{f}(0)}{\kappa_1 q_1^2} = \frac{4\pi^2 i \gamma Q_0 \tilde{f}(0)}{\omega \rho C}$$

$$= \frac{i\gamma Q_0 \pi a^2}{\omega \rho C}, \qquad (59)$$

which is independent of κ and identical to the result for a bulk material, Eq. (36).

In the one-dimensional limit $q_{1z} \to q_1$, $q_{lz} \to q_l$ and since $q_l \ll |q_1|$

$$u_{1z}(r,0) = \frac{i\gamma A_1}{q_1}(1 - R_1)f(r). \qquad (60)$$

Finally, using Eq. (48) we obtain

$$u_{1z}(r,0) = \frac{i\gamma Q_0}{\omega \rho C}f(r). \qquad (61)$$

This is a result independent of thermal conductivity and, in fact, independent of the presence of the layer and identical to the result for a bulk sample, Eq. (38). *Thus in the one-dimensional regime the presence of a layer has no effect on the thermoelastic displacement.*

For the three-dimensional case we substitute a gaussian source term, Eq. (15), into Eq. (48) and using Eq. (60) we obtain:

$$u_{1z}(r,0) = \frac{\gamma Q_0 a^2}{4\kappa_2} \int dq_r e^{-\frac{q_r a^2}{4}} \frac{J_0(q_r r)}{q_r}\left[1 + q_r d\left(\frac{\kappa_2 - \kappa_1}{k_2}\right)\right]$$

$$= \frac{\gamma Q_0 a^2}{4\kappa_2}\left\{\frac{1}{2}E_i\left(\frac{-r^2}{a^2}\right) - \ln\left(\frac{r}{a}\right)\right.$$

$$\left. + \sqrt{\pi}\left(\frac{d}{a}\right)\left(\frac{\kappa_2 - \kappa_1}{\kappa_2}\right)I_0\left(\frac{r^2}{2a^2}\right)e^{\frac{-r^2}{2a^2}}\right\} \qquad (62)$$

As in the bulk samples, u_{1z} varies as ω^{-1} in the one-dimensional regime, but is independent of ω in the three-dimensional regime.

Similarly, for the local thermoelastic slope assuming a gaussian source term, we obtain:

$$\delta(r,0) = \frac{\gamma Q_0 a}{4\kappa_2}\left\{\left(\frac{a}{r}\right)\left[1 - e^{-r^2/a^2}\right]\right.$$

$$\left. + \sqrt{\pi}\left(\frac{d}{a}\right)\left(\frac{r}{a}\right)\left(\frac{\kappa_2 - \kappa_1}{\kappa_2}\right)\left[I_0\left(\frac{r^2}{2a^2}\right) - I_1\left(\frac{r^2}{2a^2}\right)\right]e^{-r^2/2a^2}\right\} \quad (63)$$

III. DISCUSSION

Several interesting insights emerge from this analysis. First, in the one dimensional regime, where $a \gg \mu$, the surface temperature varies as $\kappa^{-1/2}$ and as $\omega^{-1/2}$, while the surface thermoelastic expansion is independent of κ but varies as ω^{-1}. In the one-dimensional regime, the surface thermoelastic expansion is completely insensitive to the presence of a thin layer ($d < a, \mu_1$) that differs from the underlying substrate only in its thermal conductivity. This is a consequence of the fact that the total displaced volume is always independent of thermal conductivity, and since in one-dimension the local themoelastic expansion must be the same everywhere, therefore the local expansion is also independent of thermal conductivity and thus insensitive to the presence of the layer. Only the surface temperature "sees" the layer. And it sees this layer as a small perturbation to the surface temperature term coming from the underlying substrate. This perturbation scales with the ratio (d/μ_1) and, in the extreme case of $\kappa_1 \ll \kappa_2$, with the ratio ($\sqrt{\kappa_2}/\sqrt{\kappa_1}$).

In the three-dimensional regime ($a \ll \mu$) the surface temperature and the surface thermoelastic expansion are both independent of ω but vary as κ^{-1}. This independence from ω necessarily leads to signals having constant phase. Thus in the three-dimensional limit the amplitude of the signal is dependent on, but the phase is totally independent of, thermal conductivity. Unlike in the one-dimensional regime, in the three-dimensional regime both the thermoelastic expansion as well as the surface temperature can see the presence of the thin layer. Both see this layer as a small perturbation to the signal from the underlying substrate. However, the perturbation to the surface temperature is usually stronger than that to the surface thermoelastic deformation. For example, in the case where $\kappa_1 \ll \kappa_2$, the perturbation to the surface temperature is $\dfrac{2}{\sqrt{\pi}}\left(\dfrac{d}{a}\right)\left(\dfrac{\kappa_2}{\kappa_1}\right)$, while the perturbation to the surface thermoelastic expansion is $\sqrt{\pi}\left(\dfrac{d}{a}\right)$.

Thus, under almost all conditions the surface temperature is a much more sensitive probe of a thin surface layer if that layer differs from the underlying substrate primarily in its thermal conductivity. Thermal wave techniques that are a measure of surface temperature, such as modulated reflectance, optical beam deflection and photothermal radiometry, will be much more sensitive probes of such thin layers than will thermal wave techniques that are a measure of local surface thermoelastic expansion, such as modulated surface displacement or modulated surface deflection.

The fact that the total displaced volume is always independent of thermal conductivity provides a convenient means of determining whether or not a layer differs from the underlying substrate primarily in its thermal properties. By mapping out the local surface displacements, one can obtain a measure of the total surface displacement. If this volume does not change appreciably with different amounts of material processing (e.g. ion implantation) then one can conclude that the process primarily changes the thermal parameters, such as thermal conductivity, with only minor changes to the elastic or other properties of the material.

The independence of both the surface temperature and thermoelastic expansion from ω in the three-dimensional limit implies that phase-monitoring experiments in this limit will be far less sensitive than in the one-dimensional regime. However, once again, one must keep in mind that only the surface temperature can provide information about a layer in the one-dimensional regime.

Finally, even though the foregoing analysis was performed for a continuous surface layer, these insights apply equally well to the case of subsurface features, since such features can be treated qualitatively as localized "layers".

IV. REFERENCES

1. A. Rosencwaig, *Photoacoustics and Photoacoustic Spectroscopy*, (John Wiley, New York, 1980).

2. *Special issue on Photoacoustic/Photothermal Phenomena:* J. Phys. (Paris) **C6-44** (1983).

3. *Special issue on Photoacoustic/Photothermal Phenomena:* Can. J. Phys. **64** (1986).

4. *Photoacoustic and Thermal Wave Phenomena in Semiconductors*, (A.

Mandelis, Ed.) (North-Holland, New York, 1987).

5. *Photoacoustic and Photothermal Phenomena*, (P. Hess and J. Pelzl, Eds.) (Springer-Verlag, Berlin, 1988).

6. *Photoacoustic and Photothermal Phenomena II*, (J.C. Murphy, J.W. Maclachlin Spicer, L.E. Aamodt and B.S.H. Royce, Eds.) (Springer-Verlag, Berlin, 1990).

7. *Principles and Perspectives of Photothermal and Photoacoustic Phenomena*, (A. Mandelis, Ed.) (Elsevier, New York, 1992).

8. A.C. Boccara, D. Fournier and J. Badoz, Appl. Phys. Lett. **36**, 130 (1980).

9. J.C. Muraphy and L.C. Aamodt, J. Appl. Phys. **54**, 481 (1983).

10. P.-E. Nordal and S.O. Kanstad, in *Scanned Image Microscopy*, (E.A. Ash, Ed.) 331 (Academic, London, 1980).

11. G. Busse, in *Scanned Image Microscopy*, (E.A. Ash, Ed.) 341 (Academic, London, 1980).

12. A. Rosencwaig, J. Opsal, W.L. Smith and D.L. Willenborg, Appl. Phys. Lett. **46**, 1013 (1985).

13. J. Opsal and A. Rosencwaig, Appl. Phys. Lett. **47**, 498 (1985).

14. S. Ameri, E.A.Ash, V. Neuman and C.R. Petts, Electron Lett. **17**, 337 (1981).

15. H.K. Wickramasinghe, Y. Martin, D.A.H. Spear and E.A. Ash, J. Phys. (Paris) **C6-44**, 191 (1983).

16. N.M. Amer, J. Phys. (Paris) **C6-44**, 185 (1983).

17. J. Opsal, A. Rosencwaig and D.L. Willenborg, Appl. Opt. **22**, 3169 (1983).

18. A. Rosencwaig and A. Gersho, J. Appl. Phys. **47**, 64 (1976).

19. L.C. Aamodt and J.C. Murphy, J. Appl. Phys. **52**, 490 (1981).

20. F.A. McDonald, J. Appl. Phys. **52**, 381 (1981).
21. J. Opsal and A. Rosencwaig, J. Appl. Phys. **53**, 4240 (1982).
22. A. Mandelis and J.D. Lymer, Appl. Spectrosc. **39**, 473 (1985).
23. H.J. Vidberg, J. Jaarinen and D.O. Riska, Can. J. Phys. **64**, 1178 (1986).
24. A. Rosencwaig and J. Opsal, IEEE Trans. UFFC-**33**, 516 (1986).

4

THERMAL WAVE MONITORING AND IMAGING OF ELECTRONIC MATERIALS AND DEVICES

Allan Rosencwaig

Therma-Wave, Inc.
Fremont, CA 94539

I.	INTRODUCTION	74
II.	EXPERIMENTAL METHODOLOGY	75
III.	ION IMPLANT MONITORING	77
IV.	ETCH MONITORING	85
V.	METALLIZATION MONITORING	86
VI.	POLYCRYSTALLINE AND AMORPHOUS SILICON	88
VII.	IMAGING OF SUBSURFACE SILICON DEFECTS	90
VIII.	IMAGING OF SUBSURFACE DEFECTS IN METAL LINES	99
IX.	CONCLUSIONS	106
X.	REFERENCES	106

I. INTRODUCTION

Thermal wave methodologies have found many diverse applications: spectroscopy, thermal and transport property evaluations, chemical and photochemical studies, and biological and medical studies [1]. One of the fastest growing areas of thermal wave research is non-destructive evaluation on a wide range of materials, the subject of this volume. In this Chapter I will focus on thermal wave investigations of semiconductor materials and devices. Since several of these studies have been discussed previously [2], I will limit the discussion to more recent work, and in particular to recent studies performed at Therma-Wave, Inc. using the modulated reflectance and modulated surface deflection techniques.

The goal of these studies has been to provide the semiconductor industry with measurement and inspection systems that can enhance the productivity of the integrated circuit manufacturing process and the quality of the finished devices. Measurement and inspection in the semiconductor field are most valuable when they can be performed directly on the product wafers during the fabrication process. Thus it is essential that the measurement method used be non-contact, non-damaging and non-contaminating, rapid and able to measure within very small geometries in the range of 1-5μm. Very few thermal wave methodologies can satisfy these demanding criteria. The gas-microphone method [1] requires a closed cell approach for good sensitivity thereby making it inconvenient and slow for the study of semiconductor wafers, which now range up to 8" in diameter. Piezoelectric [1,3] and pyroelectric [4,5] methodologies are inappropriate due to their contact nature. Infrared radiometric methods [6,7] tend to be too insensitive, difficult to quantitate because of unknown emissivities on product wafers and also cannot meet the spatial resolution criteria. Photothermal deflection techniques [8,9] would be difficult to adapt to such large objects without a sizable offset distance from the surface, thereby greatly reducing sensitivity and limiting the spatial resolution.

The most practical techniques involve the use of highly focused laser beams that both generate and detect thermal waves at the front surface of the product wafer. These techniques would then be modulated reflectance [10-12], modulated surface deflection [13,14] and modulated surface displacement [15,16]. In modulated reflectance the reflected power of the probe laser beam is modulated by the thermal wave-induced variations in the optical constants (n and k) of the sample and thus in its optical reflectance. In modulated surface deflection, the local thermoelastic deformation of the sample surface is detected by directing the probe beam to the side of the pump beam and measuring the slope of the small (typically < 1 Å) thermoelastic bump or surface distortion. In modulated surface displacement an interferometer is used to detect the height of this local modulated thermoelastic bump or surface distortion. Since it is most difficult, if not impossible, to distinguish in an interferometer between the signal due to modulated reflectance and that due to modulated surface displacement, this technique is not as useful as the previous two.

This Chapter will deal with the applications of the techniques for thermal-wave modulated reflectance, surface deflection and surface displacement in the field of electronic materials and devices. In the applications described below the thermal waves are generated by a highly focused intensity-modulated pump laser and are detected with a probe laser either through modulated reflectance or through modulated surface deflection. These applications include ion implant monitoring, etch studies, metal film analysis, studies on amorphous and polycrystalline Si, and imaging of defects in both Si substrates and in metal layers.

II. EXPERIMENTAL METHODOLOGY

The experiments described in this chapter were performed with a commercial thermal wave system [17], that utilizes the laser-induced modulated reflectance and modulated deflection techniques. Figure 1 depicts the basic optical arrangement in this system. The 488-nm beam of a 35mW Ar^+-ion laser is intensity-modulated in the 1-10 MHz frequency range with an acousto-optic modulator. The pump beam is then directed through a beam expander and focused to an ~1µm diameter spot on the sample with a sample incident power of ~10mW. The 633-nm beam of a 5-mW He-Ne laser, the probe beam, is directed through a beam expander, a polarizing beam splitter and quarter-wave plate, reflected off a dichroic mirror and focused collinearly with the Ar^+-ion pump beam onto a 1 µm diameter spot on the sample with an incident power of ~3mW. The retroreflected probe beam passes through the quarter-wave plate again, and since its polarization is now rotated 90° with respect to the incoming beam from the He-Ne laser, the retroreflected beam is directed by the polarizing beam splitter to the photodetector. The photodetector is a bi-cell that can be used in either a sum or difference mode.

When used in the sum mode the photodetector acts as a detector of total reflected power in the probe beam and thus measures only the modulated reflectance signal, that is the variations in the reflected probe beam power induced by the thermal waves (and electron-hole plasma waves). By purposely underfilling the photodetector any thermal wave induced deflections in the probe beam due to the presence of a local thermoelastic distortion are not detected. When the modulated reflectance signal is being measured, the probe He-Ne beam is usually kept coincident with the pump beam on the sample surface.

When one desires to measure the probe beam deflections induced by the presence of the thermoelastic distortions or bump, the probe beam is displaced approximately one spot size (~ 1 µm) from the pump beam and the bi-cell photodetector is operated in the difference mode. Additional optical

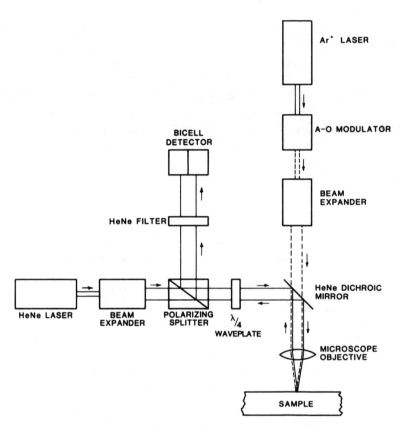

Fig. 1 Schematic depiction of key optical elements of the commercial thermal-wave apparatus. (From Ref. [17], with permission).

elements maintain the d.c. position of the probe beam at the center of the bi-cell so that only the modulated deflection is measured. Since the deflection is very small, the modulated reflectance changes in the probe beam power are cancelled out to first order and only the modulated deflection is measured. This bi-cell photodetector arrangement is thus able to measure independently both the modulated reflectance and the modulated deflection signals.

The modulated signals from the photodetector are analyzed with a phase-sensitive synchronous detection system capable of operating up to 10 MHz. Modulated reflectance changes $\Delta R/R$ as small as $10^{-7}/\sqrt{Hz}$ and modulated deflections corresponding to surface displacements as small as 10^{-4} A/$\sqrt{Hz}$ can be measured with this thermal wave system.

III. ION IMPLANT MONITORING

Control of the accuracy and uniformity of the ion implant dose is critical to good device performance and wafer yield in the manufacture of integrated circuits. This is particularly true for MOS devices where threshold voltage control is paramount since slight variations in certain electrical parameters such as the depletion drain current and the transistor threshold voltage can dramatically affect wafer yield and device speed. Most implants are performed fairly early in the IC fabrication cycle. There is thus often a 3 to 8 week time lag between the critical implant steps and the first electrical measurements that indicate whether the ion implant process was properly executed. Therefore there is a major need for an effective real-time monitor for these critical implant process steps, and in particular for a monitor on the product wafers themselves. The thermal wave system described above answers this need since it is able to measure both the implant dose and its uniformity across the wafer and to do so non-destructively on product wafers and immediately after the ion implantation process [18-22].

Of course, a thermal wave technique does not measure the ion implant dose (i.e. the concentration of implanted ions) directly. What the thermal wave technique measures is the extent of lattice disorder that has been created by the ion implantation process. This has been verified in a study where an excellent correlation was found between the thermal wave signal and the density of displaced atoms (D_{da}) as determined from Rutherford backscattering for implanted Si samples (see Fig. 2) [23]. Since the ion implantation process can produce a significant amount of damage, the thermal wave system is a sensitive monitor for this process and thus can be used to measure the dose and uniformity of the implant. The most sensitive indicator of this lattice damage is the modulated reflectance technique. This sensitivity comes about from the plasma wave contribution to the modulated reflectance signal as described in Ref. [11].

A typical correlation curve between thermal wave signal and implant dose is shown in Fig. 3 [24]. We see that the lattice damage induced by the 50 keV beam of B^+ ions causes the modulated reflectance signal to increase above that for non-implanted Si in a monatomic and pronounced manner from less than 10^{-4} $\Delta R/R$ to almost 10^{-2} $\Delta R/R$. (Note that 1 TW unit = 10^{-6} $\Delta R/R$). Of particular importance is that the damage generated by ion implantation is highly reproducible and thus the modulated reflectance method has proven to be very practical in monitoring this critical process.

We have found appreciable ion dose sensitivity to extend below 10^{10} ions/cm^2 and up to 10^{17} ions/cm^2. Of particular interest is the sensitivity at low dose. Experiments have shown that the modulated reflectance signal is more sensitive than Rutherford backscattering or Raman spectroscopy in measuring the extent of ion implant induced lattice damage and thus of the ion implant dose [25].

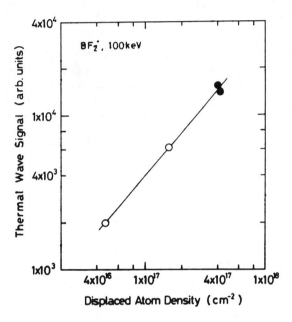

Fig. 2 Correlation of thermal wave (TW) signal intensity with displaced atom density (D_{da}) for BF_2 implantation in Si_1 where implantation was performed at 100 keV. (From Ref. [23], with permission).

Thermal wave measurements are now widely accepted as the industry standard for characterization and monitoring of ion implantation at low to medium dose levels. Until recently, however, these measurements had a practical upper limit of about 10^{15} ions/cm^2 for As and P implants. This is because implants at higher doses form a subsurface amorphous layer that causes a non-monotonic behavior in the modulated reflectance signal as shown in Fig. 4. This layer of amorphous Si is produced whenever crystalline Si is implanted at sufficiently high dose and energy. For example, arsenic implants will cause amorphization at doses greater than 10^{14} ions/cm^2 for energies above 10 keV. On the other hand phosphorous implants require higher doses and energies to cause Si amorphization. As the dose is increased above the amorphization level, local lattice damage does not increase since the amorphous lattice is completely damaged already. But rather the thickness of the amorphous layer increases with dose for a fixed energy or conversely with increases in energy for a fixed dose [26].

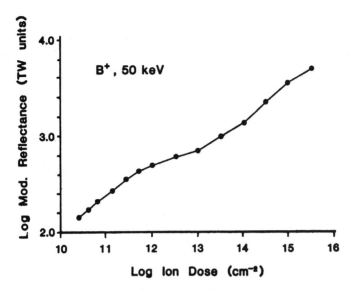

Fig. 3 Modulated reflectance signal from Si wafers vs. implanted dose of 50 keV ^{11}B$^+$ ions. Note 1 TW unit = 10^{-6} ΔR/R. (From Ref. [24], with permission).

Fig. 4 Variation of modulated reflectance thermal wave signal (TW unit = 10^{-6} ΔR/R) with ion implant dose. Note non-monotonic behavior at high dose for As and P implants due to the presence of an amorphous Si layer.

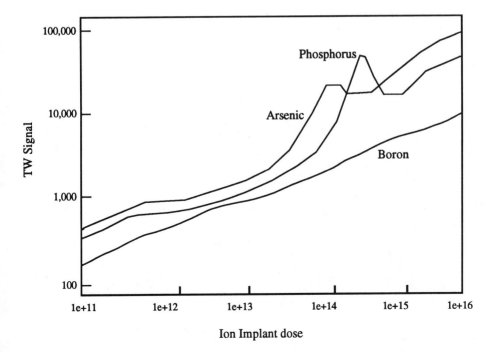

The amorphous Si layer is quite thin, typically a few hundred to a few thousand Å in thickness [26]. This means that the layer is thin not only to the thermal and plasma waves but also to the penetration depth of the He-Ne probe laser. That is, the thickness of the amorphous layer is much smaller than the thermal and plasma diffusion lengths and smaller than the He-Ne absorption length. Thus a calculation of the modulated reflectance signal must include both optical as well as plasma wave interference effects (Note that the plasma wave is by far the strongest component of the modulated reflectance signal for implanted Si as discussed in Chapter 7, Ref. [2]).

As shown in Fig. 5 these calculations [27,28] show a clear optical as well as plasma wave interference behavior with increasing amorphous layer thickness. Because of the non-monotonicity of the modulated reflectance signal, a simple correlation to dose is not possible. Therefore, in order to

Fig. 5 High dose ion implantation results in the formation of a subsurface amorphous silicon layer in crystalline silicon. The d.c. laser reflectivity (R HeNe) demonstrates the presence of optical interference while the modulated reflectance signal (TW Signal) demonstrates the presence of both optical and plasma wave interference effects.

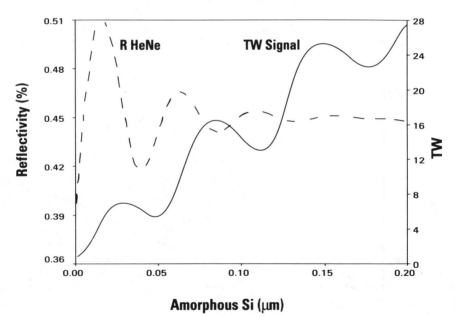

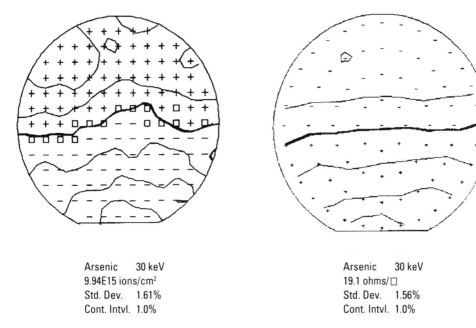

Arsenic 30 keV	Arsenic 30 keV
9.94E15 ions/cm²	19.1 ohms/□
Std. Dev. 1.61%	Std. Dev. 1.56%
Cont. Intvl. 1.0%	Cont. Intvl. 1.0%

Fig. 6 Comparison of contour maps, from the thermal wave measurements (left) and 4-point probe sheet resistance measurements (right), measured on the same implant wafer. The contours reverse since a higher dose corresponds to a lower sheet resistance value.

extract dose from the modulated reflectance and d.c. reflectivity measurements a model calculation must be performed to determine the thickness of the amorphous layer [27, 28]. Since the amorphous layer thickness is a monotonically increasing function of dose a simple calibration between the two can be made. The validity of this model can be illustrated by comparing the thermal wave results to sheet resistance measurements made by 4-point probe after the implant has been annealed as illustrated in Fig. 6. The sheet resistance method is the conventional means for high dose measurements, although it can be used only on test wafers since it is a contact technique.

Although this thermal wave system is an excellent tool for obtaining implant uniformity maps on test wafers, its primary advantage for the semiconductor industry is its ability to monitor the implant directly on the product wafers. It is able to do this because the modulated reflectance thermal wave technique is non-contact and non-damaging, is able to measure in very small geometries because of the 1 micron laser spot size, and the measurements can be done immediately after implant and through the screen oxide without any additional process steps.

Of great interest to the IC manufacturer is the correlation between the thermal wave signal obtained after the ion implantation process and the ultimate electrical characteristics of the devices at the end of the manufacturing cycle. A correlation study [29] where the thermal wave signal was correlated against both the transistor threshold voltage, and the depletion drain current showed excellent correlation between the thermal wave signal and the electrical parameters for this low dose implant application.

Another product wafer study was performed for a high dose implant. The device wafers were implanted at 5 different doses around a target dose. The implant was Arsenic with a nominal dose of 9×10^{15} ions/cm^2. No screen oxide was used. Measurements were performed on the wafers at three locations, top, center and bottom. After completing the measurements, the wafers were sent on to complete the manufacturing process. Electrical measurements were made on simple resistor test structures to determine the implanted dose. The results are plotted in Fig. 7. A clear correlation can be seen between contact resistance and thermal wave measured dose. The significant scatter of the data in contact resistance readings for each dose grouping may be attributable to critical dimension (CD) variations in the resistor test structures.

There is a group of measurements at approximately 1×10^{16} that do not correspond to a particular implant target dose. These points do, however, lie in a smooth line with the other data points with respect to contact resistance. This indicates that these measurements actually represent areas of the wafer that received a substantially different dose from the implant target.

The thermal wave system can thus be used for real-time monitoring of the critical ion implant process steps. Measurements are usually performed

Fig. 7 Correlation of contact resistance to thermal wave measured dose on device structures on product wafers.

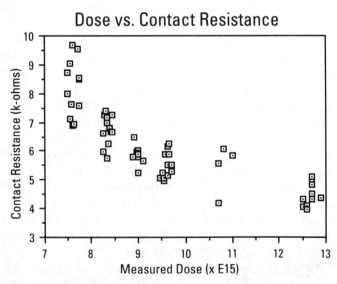

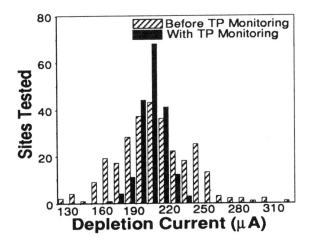

Fig. 8 Recorded variability in device depletion current before and after statistical process control with a thermal wave system (From Ref. [30], with permission).

on one or two product wafers out of every cassette and at 5 sites per wafer. The data from each wafer are tabulated and analyzed for both dose and uniformity, then plotted on process trend charts to ascertain if the implant process is within specified control limits. Deviations beyond these control limits or data with trends towards a control limit are automatically brought to the operator's attention. This statistical data analysis and data management is crucial in making the thermal wave system an effective statistical process control tool.

Finally, in Figure 8 we show how the use of a thermal wave system to control the ion implant has significantly reduced unwanted variations in device electrical performance for a customer [30]. This reduction in device performance variability significantly improved the manufacturer's yield of higher speed and thus higher revenue devices.

Of course the ability to monitor implant dose on either test or product wafers is of limited value, if such measurements cannot be performed in a highly reproducible fashion. The thermal wave system described here can provide the necessary reproducibility over long periods of time, with a long-term repeatability of 0.15% in thermal wave signal which corresponds to roughly 0.6% in dose.

In summary, the non-contact, non-destructive, and rapid nature of these modulated reflectance measurements, combined with both low-dose and high dose sensitivity and micron-scale resolution, makes this technique particularly suited to production implant monitoring of actual patterned IC product wafers as well as test wafers.

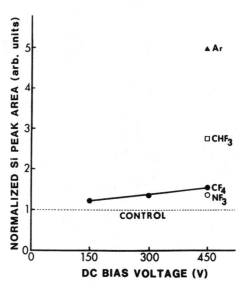

Fig. 9 Integrated silicon peak from ion channeling spectra for silicon etched under various conditions, normalized with respect to a control wafer which was not exposed to a plasma. (From Ref. [33], with permission).

Fig. 10 Modulated reflectance vs. d.c. bias for silicon etched in Ar, CF_4, NF_3, and CHF_3 plasma (1 TW unit = 10^{-6} $\Delta R/R$). (From Ref. [32], with permission).

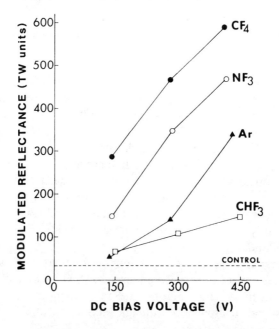

IV. ETCH MONITORING

Another application of the thermal wave system in the field of semiconductor processing is in the monitoring and evaluation of Si damage arising out of energetic dry etch processes such as reactive ion etching (RIE) and plasma etching (PE). I have described this application in some detail in an earlier review [31] and thus will not devote much attention to this topic in this review.

We have investigated this application further and in particular we have correlated thermal wave results with results obtained from transmission electron microscopy (TEM) and Rutherford Backscattering (RBS) ion channeling [32]. In this study silicon wafers were reactive ion etched using Ar, CF_4, NF_3 and CHF_3 etch gases at d.c. bias voltages ranging from 150 to 450V. The TEM and RBS techniques can give qualitative and quantitative information about the extent of lattice damage that resulted from the RIE process. The RBS data displayed in Fig. 9 show that the extent of lattice damage, as determined by the area of the Si channel peak normalized to an unetched initial sample, increases with both bias voltage and with etch chemistry. The Ar and CHF_3 etches produced the greatest lattice damage. These results also agree with the TEM data. Figure 10 shows the modulated reflectance signal obtained for these samples. The fact that the modulated reflectance, $\Delta R/R$, also increases with bias voltage suggests that it is at least a partial measure of RIE-induced lattice damage.

However, the dependence of $\Delta R/R$ on etch chemistry is different from that seen in the RBS or TEM data. This indicates that the thermal wave

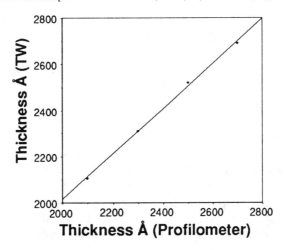

Fig. 11 Correlation of thickness of WSi metal film as determined by non-contact thermal wave method (y-axis) and contact profilometer method (x-axis). (From Ref. [30], with permission).

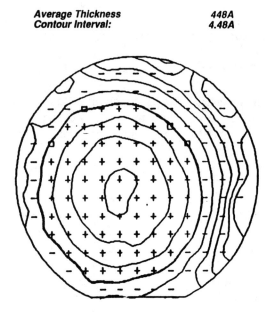

Fig. 12 Thickness uniformity map for a thin titanium film showing thermal wave sensitivity to metal thickness changes of less than 5 Angstroms. (From Ref. [30], with permission).

modulated reflectance technique is sensitive not only to the structural lattice damage that can be detected by TEM and RBS methods, but also to some other etch-induced perturbations such as contamination or surface states [33, 34] which cannot be detected with either the TEM or the RBS method.

V. METALLIZATION MONITORING

The metallization system on an IC device is of fundamental importance to its performance. Modern IC devices use several different metal films for interconnects, diffusion barriers and as input-output conductors. In controlling metallization, the process engineers continually measure four critical parameters; thickness, sheet resistance, grain size distribution and optical reflectivity. They employ a variety of instruments to make these measurements, many of which cannot be used directly on the product wafer.

Here again the thermal wave system can help provide the needed data directly from the product wafer and in a totally non-destructive and real-time fashion. All of the metal films used in the IC industry can be measured whether they be Al alloy conductors, titanium or platinum barrier metals, or tungsten silicide interconnects [30]. The thermal wave signal is sensitive to both the thickness and the thermal conductivity of the top metal layer as long

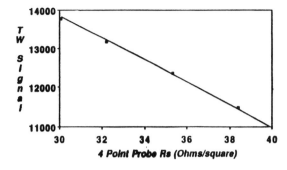

Fig. 13 Correlation of the thermal wave signal(1 TW unit = 10-6 ΔR/R) with 4-point probe sheet resistance for a WSi metal film.
(From ref. [30], with permission).

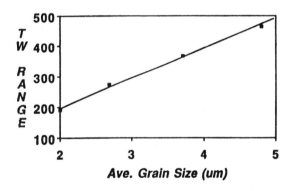

Fig. 14 Correlation of thermal wave range signal (avg. peak-to-peak variation in a line scan) with average grain size for a series of Al(Si) metal films of constant thickness.
(From Ref. [30], with permission).

as there is a reasonable difference between the thermal properties (conductivity, density, specific heat) of the film and the underlying layer or substrate. Thickness can be accurately determined from 100A to over 1 μm with a typical reproducibility of ±2%. These thickness measurements are usually performed with the system operating in the modulated deflection mode. Figure 11 shows the excellent correlation in metal film thickness obtained with the conventional contact profilometer method. Figure 12 shows a thermal wave map of thickness uniformity for a thin titanium film with a thickness sensitivity of less than 5 Angstroms.

Figure 13 shows the correlation between thermal wave signal and sheet resistance as determined from a contact 4-point probe. These two signals are, as expected, well correlated since the thermal conductivity in a metal is directly related to its electrical conductivity by the Wiedemann-Franz law.

The thermal wave signal is also very sensitive to the presence of boundaries whether they be inter-layer or inter-grain, because of thermal wave reflection and scattering from these boundaries. Thus line scans of this signal over a metal film provide information about grain size distribution. This is shown in Figure 14 where we see a linear relationship between thermal wave signal and average grain distribution in an Al(Si) alloy. The average grain distributions were determined in the conventional manner from post-etch SEM micrographs.

Finally highly precise optical reflectivity measurements can be made of the metal surfaces at the pump and probe laser wavelengths of 488 and 633nm, respectively.

All of these measurements, which currently have to be done using a number of different instruments, and generally on test wafers only, can now be performed non-destructively, and with micron scale resolution directly on the product wafers with the thermal wave system. This real-time product wafer capability permits use of the thermal wave system for effective statistical process control of the metallization process in the same manner as in the ion implant application.

VI. POLYCRYSTALLINE AND AMORPHOUS SILICON

The sensitivity of the thermal wave signal to the presence of grain boundaries is particularly high in the case of polycrystalline and amorphous Si films deposited on Si wafers [35]. This is illustrated in Figs. 15 and 16. Both figures demonstrate that the $\Delta R/R$ signal of these Si structures (ranging from 2×10^{-3} to 3×10^{-2} $\Delta R/R$) is substantially higher than that observed from typical chemomechanically polished monocrystalline Si wafers where the $\Delta R/R$ signal ranges from 3×10^{-5} to 2×10^{-4} $\Delta R/R$. (Note that 1 TW unit = 10^{-6} $\Delta R/R$). This large increase is signal is primarily a result of the much lower ambipolar diffusion coefficient for the plasma waves because of the presence of the many grain boundaries in the polycrystalline and amorphous silicon.

For the amorphous Si/monocrystalline Si structures the $\Delta R/R$ signal increases with an increase in the thickness of the amorphous Si film, while for the polycrystalline Si/monocrystalline Si structure it decreases with an increase in the thickness of the polycrystalline film. The difference can be explained, if one postulates that the $\Delta R/R$ signal responds in a monotonically increasing manner to the increasing total grain boundary area within the probed volume (~1 μm^3). For the amorphous Si film sample, the total grain boundary area within the probed volume will increase monotonically with increasing film thickness (at least up to 1 μm thickness), since the average grain size of amorphous Si is independent of film thickness. However in a polycrystalline Si film the average grain size increases with increasing film thickness and thus the total grain boundary area will decrease with increasing film thickness. This concept is further verified by noting that the average grain size of polycrystalline Si will also increase with thermal annealing time and thus that the $\Delta R/R$ signal should decrease with thermal annealing time. This is indeed the case as seen in Fig. 17.

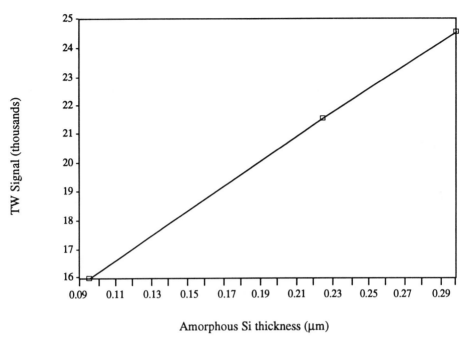

Fig. 15 Variation of TW signal with respect to amorphous Si film thickness. (From Ref. [35], with permission).

Fig. 16 Variation of TW signal with respect to undoped poly Si film thickness. (From Ref. [35], with permission).

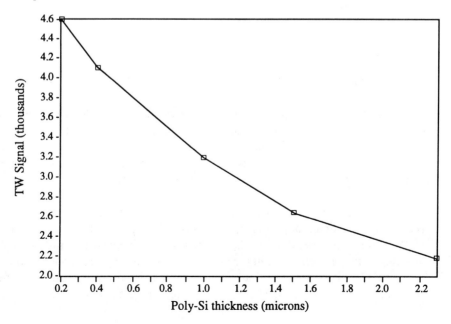

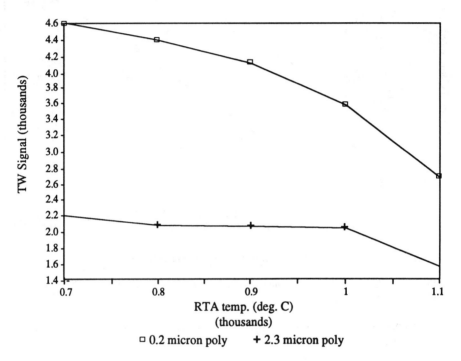

Fig. 17 Variation of TW signal of undoped poly Si film with respect to RTA temperature (°C). (From Ref. [35], with permission).

VII. IMAGING OF SUBSURFACE Si DEFECTS

The IC industry has improved process yield and device reliability considerably over the last few years. Much of the improvement is the result of continuing reduction in the defect density on IC wafers, a result made possible in part through the use of suitable defect detection and wafer inspection systems. These systems, however, are useful only for surface or visible defects.

A second major category of defects that is, as yet, largely unaddressed is that of nonvisible and subsurface defects. Examples of these defects are: (i) microstructural defects in silicon such as precipitates, dislocations and stacking faults, defects that can seriously degrade device performance through emitter-collector shorts, junction leakage and carrier lifetime reduction; and (ii) process-induced voids, notches and microcracks in metal lines, defects which can lead to serious reliability problems with VLSI devices. These non-visible silicon and metal defects can degrade device yield in both a continuing yield loss manner and with periodic production-halting crashes. Of perhaps greater economic consequence, however, is the damage to end user systems resulting from failures in the field. The cost of such failures in the field to the IC manufacturer often far exceeds the cost of the failed components themselves.

We have recently developed an inspection system to provide the needed real-time non-destructive inspection capability to address the critical issue of subsurface silicon and metal defects.

The imaging system used in these experiments is a thermal wave modulated reflectance/deflection instrument similar to the one described before. In this imaging system the sample is rapidly scanned in an x-y raster beneath the stationary pump and probe laser beams in 0.2µm steps (see Fig. 18). The beam spot sizes in the imaging system are ~0.5µm through the use of a microscope objective with higher numerical aperture. Images of 100 × 100µm areas are obtained in about 100 sec. Four different types of images are obtained with this system. Conventional scanning laser microscope (i.e. optical) images are obtained at the 633nm and the 488nm wavelengths of the probe and pump lasers, respectively. In addition, modulated reflection and modulated deflection thermal wave images are also generated. These 512 × 512 pixel images are stored in the system computer, displayed on a high resolution color monitor and can be printed out on a video printer.

Metal precipitates, dislocations and stacking faults are all detectable with this imaging system when operated in a modulated reflectance mode [36-39]. In this mode the signal is primarily determined by the local pump-generated plasma waves [11], and the defects become visible through the presence of a locally enhanced electron-hole plasma density due to the reduced plasma diffusivity and carrier recombination lifetime in the vicinity of the silicon defects. This is illustrated in Figure 19. Figure 19(a) is the modulated reflectance image of a 100 × 100 µm region of a silicon wafer where we note the appearance of two bright features between two curved fiducial marks. Figure 19(b) shows an optical image of this same region with no evidence of these two bright features. These two thermal wave features representing increased thermal wave signal tail toward each other. The sample was then subjected to oxide strip etch followed by a Wright decorative etch to make the defect optically visible. Figure 19(c) is the optical image of the same region after the etching process and shows a 40µm long stacking fault in the same location as the bright features seen in the thermal wave image. The bright features in the thermal wave image of Figure 19(a) result from the enhanced plasma density in the vicinity of the Frank partial dislocation bounding the stacking fault. This bounding dislocation starts at the silicon-oxide interface, tails down along the <111> oxide plane beneath

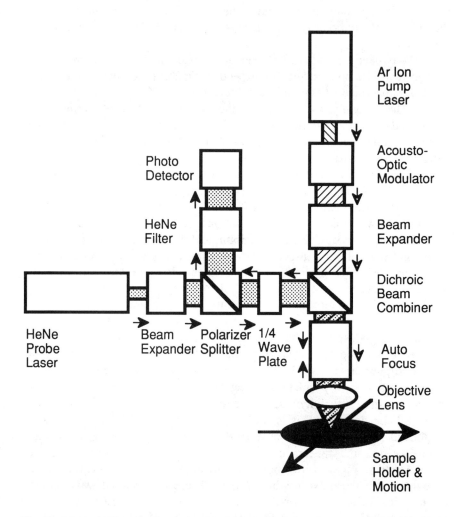

Fig. 18 Schematic diagram of optical train of thermal wave imaging system. (From Ref. [43], with permission).

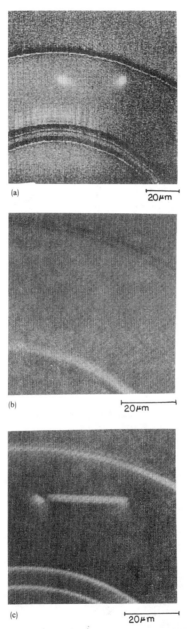

Fig. 19 (a) A thermal wave image of a 100 × 100 μm region of Si between curved fiducial marks. This image shows the near surface ends of the dislocation bounding a 40 μm stacking fault; (b) An optical image of the same region with no dislocation features in view; (c) An optical image of the stacking fault after oxide stripping and decorative etching. (From Ref [36], with permission).

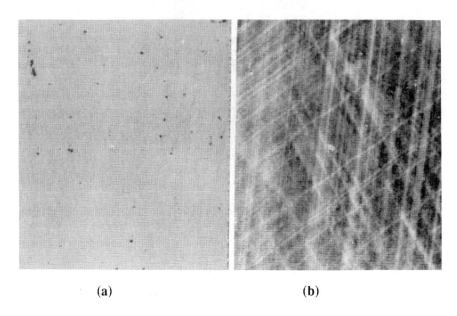

(a) (b)

Fig. 20 (a) Optical image (100×100μmm) of a mechanically stressed Si wafer; (b) Modulated reflectance image of same region showing dislocations preferentially ordered along crystallographic slip planes. (From Ref. [40], with permission).

Fig. 21 TW contour maps for Cu backside diffused Si wafers after rapid thermal processing at 1150°C for (a) 2 s. (b) 5 s. (c) 10 s. and (d) 30 s. (From Ref. [38], with permission).

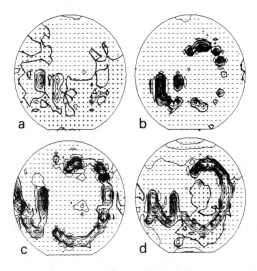

the silicon surface to a depth of about 15μm and then comes back up to the surface some 40μm away. In Figure 19(a) only the two ends of the bounding dislocation within 5μm of the surface are seen in the thermal wave image, since this is the effective penetration depth of the plasma waves generated at 1 MHz with our sub-micron focused pump beam.

The sensitivity of the modulated reflectance method to the presence of dislocations is illustrated in Fig. 20 [45]. Figure 20 (a) is a 100×100 μm optical image of a virgin Si wafer that has been mechanically stressed to produce a dislocation density of ~10^{12}/cm^3. There is no evidence of these dislocations in the optical image. Figure 20 (b) is the thermal wave image of the same region and clearly shows these dislocations preferentially ordered along the slip planes of the silicon crystal.

Metal precipitates have also been detected and imaged with this thermal wave imaging system using the modulated reflectance signal [37,38]. These metal precipitates often come from some backside metal contamination which rapidly diffuses toward the front surface during a thermal process. This is illustrated in Figures 21 and 22 which show in the thermal wave maps of the front surface the evolution of a metal precipitate signal that came about by scraping the back of the wafers with a Cu or Ni wire and then subjecting the wafers to a rapid thermal process (RTP). Figures 23 and 24 show high magnification thermal wave images of the actual Cu and Ni metal precipitates formed on the front side of the wafers.

Fig. 22 TW contour maps for Ni backside diffused Si wafers after rapid thermal processing at (a) 900°C. (b) 1000°C. (c) 1100°C. and (d) 1200°C for 10 s. (From Ref. [38], with permission).

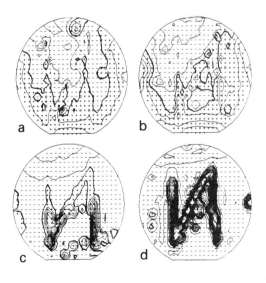

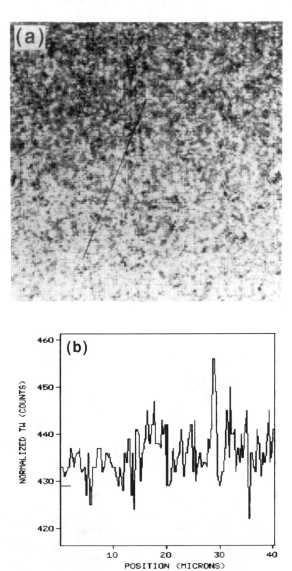

Fig. 23 (a) TW defect image pattern and (b) line plot (which shows TW signal variation between two selected points inside the map) for Cu-induced surface defect region (backside contaminated and RTA cycle at 1150°C for 30 s). (From Ref. [38], with permission).

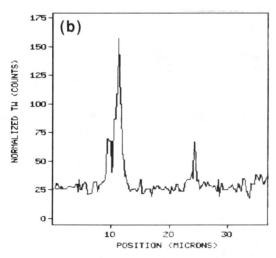

Fig. 24 (a) TW defect image pattern and (b) line plot for Ni-induced surface defect region (backside contaminated and RTA cycle at 1200°C for 10s.) (From Ref. [38], with permission).

[36]. We illustrate this application in Figure 25. In Figure 25 (a) we show an optical image of a 100 × 100μm² region of an IC trench-isolated device. Since this wafer has not been subjected to decorative etch, no silicon defects are optically visible, although electrical tests have indicated poor device performance. The thermal wave image, Figure 25 (b) clearly shows that this poor electrical performance is the result of the presence of a large number of dislocations emanating from the side walls of the isolation trenches.

The IC manufacturer can thus use this thermal wave imaging system to detect the onset of an unacceptable density of silicon defects at any stage during his manufacturing cycle in a rapid and non-destructive fashion. By early detection of a defect producing situation he will be able to correct the processing problem in a timely fashion and thereby maintain his yield.

VIII. IMAGING OF SUBSURFACE DEFECTS IN METAL LINES

Reliability of increasingly complex metallization systems is an important subject for VLSI and ULSI progress. It is presently understood that stress-induced aluminum migration, in addition to electromigration, must be better understood and controlled in future metallization systems. Driven by

Fig. 25 (a) An optical image of a 100 × 100μm² region of a VLSI bipolar, trench-isolated device. No silicon defects are visible; (b) A thermal wave image of the same region showing many sidewall dislocations defects emanating from the isolation trenches. (From Ref. [42], with permission).

(a)　　　　　　　　　　　　　　　(b)

thermal expansion mismatch between Al and surrounding materials (Si, SiO_2, etc.), the Al films are cycled between states of high tensile and high compressive stress during processing steps. Even in field use, the Al films on the completed ICs are generally under tensile stress, modulated by on-off power cycling. Aluminum migrates away from regions of high tensile stress, leaving behind vacancies. Coalescence of these vacancies generates microvoids and microcracks in the aluminum. Eventually, lines may be restricted or opened by voids or cracks, causing high resistance or device failure. Reduction or delay of voiding can be accomplished by several approaches such as adding Cu or Ti, using non-Al interlayers, or reducing passivation stress. However, because of the interplay between these and numerous related effects (Si precipitate formation, hillock formation, local variation in stress from patterned passivation layers, corrosion, etc.), selecting or modifying a metallization system is a complex, time consuming operation.

A limitation to the understanding of stress-induced and electromigration-induced voiding is presented by the tools with which one can quickly and conveniently observe the voids, notches, and other defects. Optical inspection is frequently unproductive on very narrow lines and cannot be used to detect defects beneath opaque cap metal layers. The usual method of study requires destructive stripping of passivation and cap layers so that examination in the SEM can be done. The drawbacks of this routine are that stripping is destructive, can introduce etch-related artifacts, and is slow and operator-intensive. Furthermore, SEM analysis cannot see voids that are in the interior of the metal lines unless backscatter mode is employed, and even then, typically, only to a depth of 1 kÅ to 3 kÅ.

We have reported the use of thermal wave modulated reflectance imaging to detect voids, notches and other defects in fine Al alloy lines without the drawbacks mentioned above [43,44]. Figure 26 schematically depicts the thermal wave contrast mechanism for detecting defects beneath the surface of metal lines.

We begin the discussion of results with Fig. 27, which shows a set of patterned metallization lines of different width on a sample that had received some thermal stressing. The narrowest lines are approximately 1.8 μm wide. The patterned lines are composed of approximately 2 μm of passivation materials, a cap metal layer, an Al(Si) layer of approximately 1 μm thickness, and a barrier metal layer, all on a blanket SiO_2 layer over the Si substrate. The image in Fig. 27(a) is optical and no defects are visible.

Figure 27(b) is a thermal wave (TW) image made at the same location on the same sample just described (the magnification is slightly higher in Fig. 27(b)). No layers have been removed. As will be proven below with SEM comparison, the bright objects in Fig. 27(b) are subsurface voids and notches in the Al(Si) material. These voids were invisible in Fig. 27(a) due to the presence of the opaque cap metal. Evident in Fig. 27(b) are a large number

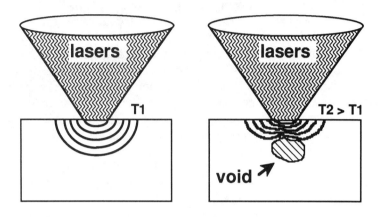

metal line sections

Fig. 26 Propagation of laser-induced thermal waves and resulting surface temperature are affected by subsurface inhomogeneities. (From Ref. [43], with permission).

of small voids associated primarily with the line edges and a few defects of larger size.

The images shown in Figs. 27(c) and 27(d) were made after subjecting this sample (still unstripped) to some additional thermal stressing - 2 hours and 12 hours of 250°C temperature, respectively. This sequence was performed to document the evolution of these nonvisible, subsurface defects under thermal stress, using thermal wave imaging. In Fig. 27(d), there is evidence that the number of small voids is decreasing, while the large defects show slight growth. Also interesting is the observation that at least one defect (the next-to-largest defect in the left, bottom quadrant of Fig. 27(b)) has disappeared in Fig. 27(d). This could be due to filling-in of a void (grain regrowth) or recovery of a delamination site.

The image in Fig. 27(d) shows clearly that most of the small voids are no longer present. The large voids have grown significantly compared to Fig. 27(b). A large void (shown with cursor line extending through it) surprisingly grew from very small size to dissect the line entirely. Figure 27(f) is the TW data plot along the cursor line in Fig. 27(d). The general behavior shown by these defects is migration of the smaller defects and reduction in their number. Since mass is conserved in this experiment (no evaporation), diffusion and coalescence of the small voids to form larger voids is expected, consistent with models published in the literature [45]. No

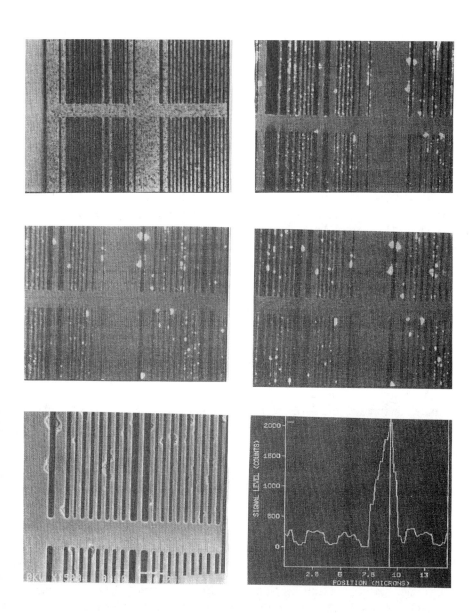

Fig. 27 Sequence of images to study migration and growth of subsurface voids in a system of passivated, cap metal-covered Al lines: the narrowest lines are approximately 1.8 μm wide. (From Ref. [43], with permission).

net motion of the largest voids is observed. A visible image made at the end of the 12-hour stress was indistinguishable from Fig. 27(a). Hence, all the void dynamics recorded in Figs. 27(b) - (d) were nonvisible.

Fig. 27(e) is a SEM micrograph of the upper right quadrant of Fig. 27(d) made after the above stress experiment was completed. The layers on top of the Al(Si) were stripped away to allow conventional SEM imaging. The one-to-one correspondence between the voids non-destructively imaged in Figs. 27(b) - (d) and the SEM post-strip image is evident. The breaking of the line by the unusually fast growing void is also confirmed by Fig. 27(e).

At least two defects are evident in Fig. 27(d) that do not appear in the SEM image. That is not unexpected, since defects located below the upper ~1000 Å of Al(Si) will not be detected by SEM; while defects in the entire thickness of the Al(Si) line will be detected in the TW image. Each TW image shown in Fig. 27 took 80 s to record. User-entered threshold TW values were entered into the job file for TW images shown here to enable analysis software to automatically identify defects and give for each the location, width, length, area, and TW signal (related to the defect "thickness"). Overall defect density values were generated, useful for process reliability monitoring.

Of particular interest is the ability to obtain quantitative data on stress-induced void growth in a totally non-destructive manner. An example of such a study is shown in Fig. 28. The average rate of growth of voids in an Al line under thermal stress that is observed in the thermal wave study is in good agreement with a diffusive void growth model [41].

Fig. 29(a) is a visible image showing a bond pad region of Al(Si, Cu) with oxide passivation. The arc-shaped line drawn onto this image encloses five dark defects. The TW image (Fig. 29(b)) made on this same area (without stripping) also shows these dark features, but further shows bright features at three of the dark defects. By comparing Fig. 29(c) (SEM photo made after stripping the oxide), we confirm that the five defects involve hillocks. A close-up SEM photo of one hillock (identified by arrow) is given in Fig. 29(d). All the dark features in Fig. 29(b) are one-to-one correlated with large and small hillocks shown in the SEM photos. However, the bright features appear only in the TW image. No interpretation of the optical or SEM photos is able to predict the observed locations of the bright features in the TW image. This is consistent with the interpretation of the bright features as subsurface inhomogeneities. We have since determined that the bright features in Fig. 29(b) are Si precipitates in the interior of the Al lines.

The last example to be discussed involves not Al-based metallization, but rather selective chemical vapor deposition tungsten (CVD W) filling of contact holes ("plugs"). For this technology, a critical aspect for low-resistance contacts is the achievement of good nucleation of the CVD W at the silicon interface. Chemical contamination from the contact hole formation

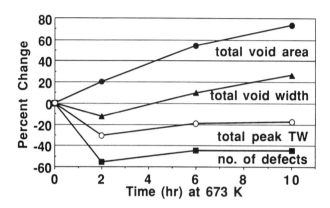

Fig. 28 Plot of major defect growth parameters (number, cumulative defect area, cumulative defect width and cumulative TW signal) for all the voids found at different steps within the selected analysis area. Each parameter is normalized to unity at 1-hr stress time. (From Ref. [44], with permission).

processing, for example, can cause loss of contact integrity. The area to be examined (Fig. 30) is one of contact "strings". The TW image in Fig. 30(b) shows a set of short metal line segments (each about 4 μm long). At the end of each segment, a plug connects to the lower conductor system. The TW image in Fig. 30(b) indicates good quality of the metallization; only one bright defect is shown. In contrast, the TW image on the same structure from a different wafer (Fig. 30(c)) shows a large number of bright features, located predominantly at the ends of the line segments at the contact plug sites. This result correlates well with measurements made by the IC manufacturer prior to the TW imaging of high-resistance or open contact strings on this die.

The peak-to-background TW signal ratio associated with detection of the imperfect contact plugs is shown in Figs. 30(d) and 30(e). A cursor line is drawn in Fig. 30(d) along a line containing two contact defects. A plot of TW signal along this line and including these two defects is shown in Fig. 30(e). The TW signal level on normal contact lines on this sample is typically 250. The "smaller" contact defect shows a peak TW signal of ~2600 and the larger, ~9000. The reproducibility of the TW signal in the normal contact areas of this sample is approximately 1% (of 250) or better.

It should be noted that the thermal wave method achieved high signal-

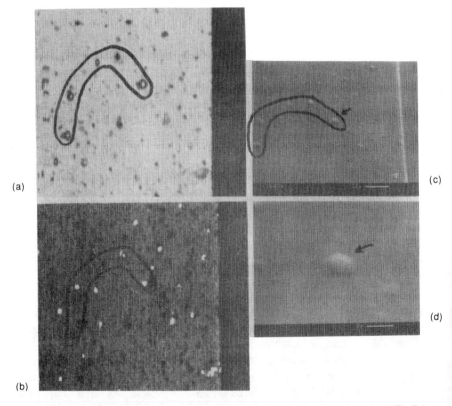

Fig. 29 Sequence of images made at same location on bond pad area of passivated Al (Si, Cu) with no cap metal. (a) optical image; (b) TW image without stripping passivation; (c) stripped-back SEM image confirming hillocks; (d) magnified SEM photo of hillock (arrowed) which has an associated bright subsurface defect as shown in (b). (From Ref. [43], with permission).

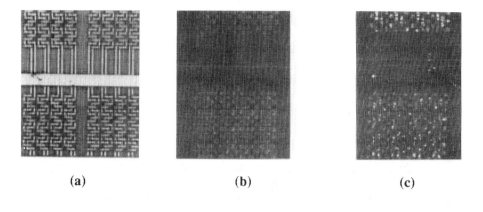

Fig. 30 Images made on selective CVD tungsten plug system. (a) optical image; (b) TW image on region of good contacts; (c) TW image on region of contacts showing bright defect features; (d) TW image showing cursor line position; (e) line plot of TW data, along cursor line in, (d) showing two defects. (From Ref. [43], with permission).

to-noise detection of these W-Si interfacial defects even though they were located 2 μm beneath the metal surface.

The cause of the high resistance was stated as erratic CVD W nucleation at the Si interface. The capability to immediately and non-destructively detect loss of selective CVD W plug integrity, as demonstrated in Figs. 30(b) and 30(c), is important due to the potential significance if CVD W plugs for future VLSI and ULSI metallization.

Thermal wave modulated reflectance imaging thus appears to be a valuable non-destructive, real-time method for inspecting and analyzing metallization integrity. Applications include the detection of subsurface voids, and the study of migration and stress-induced growth of voids. Additional applications include: defect imaging on Al(Si, Cu) pads, on multi-level metal finished die, and on selective CVD tungsten plugs. Much of these applications can be performed by other techniques only at the expense of 30x to 300x more analysis time and/or destruction of the sample. As a non-destructive method usable on wafers at different process steps, on completed dies or on aged, field-returned ICs, this thermal wave method shows promise as an important tool for use in improving metallization reliability.

IX. CONCLUSIONS

In this Chapter I have attempted to give a brief description of recent developments in the study of electronic materials and devices by thermal wave techniques that employ modulated reflectance or modulated surface deflection methods. Thermal wave technology has become a well-accepted and standard means for performing a variety of important measurement and inspection functions during the integrated circuit fabrication process. Although the requirements of the semiconductor industry are very stringent, thermal wave technology has been able to meet them. The challenge in the next few years will be to refine the current techniques and to develop new ones that will meet the even more rigorous requirements posed by the semiconductor processes of the future.

X. REFERENCES

1. See for example, A. Rosencwaig *Photoacoustics and Photoacoustic Spectroscopy*, Wiley, New York 1980; and *Principles and Perspectives of Photothermal and Photoacoustic Phenomena* (A. Mandelis, Ed.) (Elsevier, New York, 1992).

2. A. Mandelis, Ed., *Photoacoustic and Thermal Wave Phenomena in Semiconductors*, (North-Holland, New York, 1987).

3. A. Rosewncwaig and J. Opsal, IEEE Trans. UFFC-**33**, 516 (1986).

4. H. Coufal, IEEE Trans. UFFC-**33**, 507 (1986).

5. H. Coufal and A. Mandelis, in *Photoacoustic and Thermal Wave Phenomena in Semiconductors* (A. Mandelis, Ed.) 149 (North-Holland, New York, 1987).

6. S.O. Kanstad and P.E. Nordal, Can. J. Phys. **64**, 1155 (1986).

7. L.D. Favro, T. Ahmed, H.S. Lin, P. Chen, P.K. Kuo and R.L. Thomas, in *Photoacoustic and Photothermal Phenomena II* (J.C. Murphy, J.W. Machlachlan-Spicer, L.C. Aamodt and B.S.H. Royce, Eds.) 490 (Springer-Verlag, Berlin, 1990).

8. A.C. Boccara, D. Fournier and J. Badoz, Appl. Phys. Lett **36**, 130 (1980).

9. L.C. Aamodt and J.C. Murphy, J. Appl. Phys. **54**, 581 (1983).

10. A. Rosencwaig, J. Opsal , W.L. Smith and D.L. Willenborg, Appl. Phys. Lett. **46**, 1013 (1985).

11. J. Opsal and A. Rosencwaig, Appl. Phys. Lett. **47**, 498 (1985).

12. D. Guidotti and H.M. Van Driel, Appl. Phys. Lett. **47**, 584 (1985).

13. N.M. Amer, J. Phys. (Paris) **C6-44**, 185 (1983).

14. J. Opsal, A. Rosencwaig and D.L. Willenborg, Appl. Opt. **22**, 3169 (1983).

15. S. Ameri, E.A. Ash, V. Neuman and C.R. Petts, Electron. Lett. **17**, 337 (1981).

16. H.K. Wickramasinghe, Y. Martin, D.A.H. Spear and E.A. Ash, J. Phys. (Paris) **C6-44** 191 (1983).

17. The commercial thermal wave systems are products of Therma-Wave, Inc., Fremont, CA 94539.

18. W.L. Smith, A. Rosencwaig and D.L. Willenborg, Appl. Phys. Lett. **47**, 584 (1985).

19. W.L. Smith, R.A. Powell and J.D. Woodhouse, Proc. SPIE **530**, 188 (1985).

20. W.L. Smith, M.W. Taylor and J. Schuur, Proc. SPIE **530**, 201 (1985).

21. J. Opsal, M.W. Taylor, W.L. Smith and A. Rosencwaig, J. Appl. Phys. **61**, 240 (1986).

22. L.A. Vitkin, C. Christofides and A. Mandelis, Appl. Phys Lett. **54**, 2392 (1989).

23. T. Hara, S. Takahashi, H. Hagiwara, J. Hiyoshi, W.L. Smith, C. Welles, S.K. Hahn, L. Larson and C.C.D. Wong, Appl. Phys. Lett. **55**, 1315 (1989).

24. A. Rosencwaig, Mat. Res. Soc. Symp. Proc. **69**, 111 (1986).

25. T. Hara, H. Hagiwara, R. Ichikawa, S. Nakashima, K. Mizoguchi, W.L. Smith, C. Welles, S.K. Hahn and L. Larson, IEEE Electron Device Lett. **11**, 485, (1990).

26. See for example, S.M. Sze, *Physics of Semiconductor Devices, 2nd ed.*, Ch. 2 (Wiley, New York, 1981).

27. J. Opsal in *Rev. Progr. Quant. NDE* **8**, (D.O. Thompson and D.E. Chimenti, Eds.) 1241 (Plenum, New York, 1987).

28. S. Wurm, P. Alpern, D. Savignac and R. Kakoschke, Appl. Phys. **A47**, 147 (1988).

29. M.A. Wendman and W.L. Smith, Nucl. Instr. and Meth. **B21**, 559 (1987).

30. A. Rosencwaig in *Rev. Progr. Quant. NDE* **9**, (D.O. Thompson and D.E. Chimenti, Eds.) 2031 (Plenum, New York, 1990).

31. A. Rosencwaig in *Photoacoustic and Thermal Wave Phenomena in Semiconductors* (A. Mandelis, Ed.) (North-Holland, New York, 1987), Chap. 7.

32. I-W.H. Connick, A. Bhattacharyva, K.N. Ritz and W.L. Smith, J. Appl. Phys. **64**, 2059 (1988).

33. S.W. Pang, D.D. Rathman, D.J. Silversmith, R.W. Mountain and P.D. DeGraff, J. Appl. Phys. **54**, 3272 (1983).

34. A. Rosencwaig, J. Opsal, D.L. Willenborg, P. Geraghty and W.L. Smith, in *Photoacoustic and Photothermal Phenomena* (P. Hess and J. Pelzl, Eds.) 229 (Springer-Verlag, Berlin, 1988).

35. S. Hahn, W.L. Smith, C.B. Yarling, D.T. Hodul and J.-G. Park, Samsung Electronics Semiconductor Tech. J. **5**, 160 (1990).

36. B. Witowski, W.L. Smith and D.L. Willenborg, Appl. Phys. Lett. **52**, 640 (1988).

37. P. Alpern, W. Bergholz and R. Kakoschke, J. Electrochem. Soc. **136**, 3841 (1989).

38. S. Hahn, W.L. Smith, H. Suga, R. Meinecke, R.R. Kola and G.A. Rozgonyi, J. Cryst. Growth, **103**, 206 (1990).

39. W.L. Smith, C.G. Welles and A. Rosencwaig in *Rev. Progr. Quant. NDE* **9**, (D.O. Thompson and D.E. Chimenti, Eds.) 1087 (Plenum, New York, 1990).

40. A. Rosencwaig, J. Opsal, W.L. Smith and D.L. Willenborg in *Rev. Progr. Quant. NDE* **8**, (D.O. Thompson and D.E. Chimenti, Eds.) 1195 (Plenum, New York, 1989).

41. J. Bailey, E.R. Weber and J. Opsal, J. Cryst. Growth **103**, 217 (1990).

42. W.L. Smith, C.G. Welles and A. Rosencwaig in *Photoacoustic and Photothermal Phenomena II* (J.C. Murphy, J.W. Machlachlan-Spicer, L.C. Aamodt and B.S.H. Royce, Eds.) 146 (Springer-Verlag, Berlin, 1990).

43. W.L. Smith, C.G. Welles, D.L. Willenborg and A. Rosencwaig in *Tech. Proc. Semicon Japan Osaka 1989*, SEMI, Santa Clara, 1989.

44. W.L. Smith, C.G. Welles, A. Bivas, F.G. Yost and J.E. Campbell, *Proc. 1990 IRPS*, 200 (IEEE, New York, 1990).

45. F.G. Yost, D.E. Amos and A.D. Romig, *Proc. 1989 IRPS* 193 (IEEE, New York, 1989).

46. F.G. Yost, Scripta Metallurgica, **23**, 1323 (1989).

PHOTOTHERMAL RADIOMETRY OF SEMICONDUCTORS

Stephen Sheard and Mike Somekh

*University of Oxford, Department of Engineering Science,
Parks Road, Oxford OX1 3PJ, U.K.*

*University of Nottingham,
Department of Electrical and Electronical Engineering,
University Park, Nottingham NG7 2RD, U.K.*

I.	INTRODUCTION	112
	1. Photothermal Radiometry of Thermal Properties	112
	2. Photothermal Radiometry of Induced Carrier Properties	113
	3. Some Applications of Photothermal Radiometry for Semiconductor Analysis	114
II.	PTR SIGNAL GENERATION	116
	1. Generation of Photoexcited Carrier Waves	116
	2. Generation of Thermal Waves	119
	3. Black Body Emission Mechanisms	120
	4. Three Dimensional Extension	125
III.	THE PTR MICROSCOPE	126
	1. 1-D Versus 3-D Arrangement	126
	2. Configuration and Performance	128
IV.	SEMICONDUCTOR ASSESSMENT	129
	1. Direct Determination of Carrier Lifetime	129
	2. Ion Implantation Study	133
	3. Effect of Surface Recombination	138
	4. Effect of Varying Back Surface Emissivity	140
	5. Imaging of Silicon Membranes	144
	6. Imaging of Cadmium Mercury Telluride	146
V.	SUMMARY	147
VI.	REFERENCES	149

I. INTRODUCTION

Photothermal techniques rely on the generation of heat by a pulsed or modulated light source and detection of the resulting heating by one of many possible configurations. This chapter will describe the applications of radiometric detection to the non-destructive characterization of semiconductors.

Radiometric detection depends on the production of excess black body radiation whose spectrum and intensity depends on the absolute temperature of the sample. This is a fundamental phenomenon that occurs on all samples so that a variation in the absolute temperature results in a change in the black body emission emerging from the sample.

Photothermal radiometric responses can be broadly divided into (i) the examination of samples where the excess black body radiation produced arises from the temperature rise of the sample and (ii) examination of samples where the temperature rise is of secondary importance and the major source of excess black body radiation arises from the photo-induced carriers.

The situation described in (i) is the situation that pertains for non-semiconductors. Furthermore, the behavior of some extrinsic and low bandgap semiconductors having high background carrier concentrations can be explained by the thermal rather than the carrier properties of the sample. Room temperature Cadmium Mercury Telluride provides such an example and is discussed in section IV.6. The situation in (ii) applies to the examination of most semiconductor materials as described in detail in sections IV.1-4.

1. Photothermal Radiometry of Thermal Properties

The application of photothermal radiometry (PTR) to non-semiconductors where the thermal properties are probed is described in detail in Chapter 7 of this volume [1]. Nevertheless it is still useful to review some aspects of thermal wave imaging with radiometry in order to place the carrier mechanisms in perspective and to explain those results where recourse to thermal wave mechanisms is necessary (see section IV.5 and IV.6).

Photothermal radiometry of non-semiconductors relies on the fact that the emission of black body radiation increases as the fourth power of the absolute temperature according to the well known Stefan-Boltzmann Law.

$$P = \varepsilon \sigma T^4 \qquad (1)$$

where P is the radiative power emitted per unit area (Wm^{-2}), ε the emissivity, σ is Stefan's constant (Wm^{-2}K^{-4}) and T is the absolute temperature (K).

An increase in temperature by heating the sample surface thus results in an increase in the black body flux leaving the sample. The heating source can be either pulsed or continuously modulated. We are normally concerned with changes in temperature ΔT, which are small compared to the absolute temperature, so that a sinusoidal heating source will produce a sinusoidal temperature change, which in turn, produces a sinusoidal change in the black body heat flux leaving the sample. The detected signal will thus also vary sinusoidally at the frequency of the modulated heat flux. Details of the detector system are given in section III. This chapter will concentrate on the use of periodic heat sources, but, the effects of pulsed generation can be considered as a summation of the appropriate sinusoidal sources, provided, of course, the considerations mentioned above still apply.

There are several advantages of radiometric detection for non-destructive testing, many of which are discussed in this volume. One of the most important merits of radiometry from the point of view of quantitative non-destructive testing is that for non-semiconductors the signal level depends on the variation of the temperature only. This makes quantitative analysis comparatively straightforward. By solving the heat flow equation (section II.2) alone the response of the system to various samples may be predicted. The directness of the measurement variation thus provides convenient ways of measuring thermal diffusion or layer thickness [2,3].

This method contrasts with other detection schemes where the coupling of the temperature with the measured parameter complicates the quantitative analysis of the technique. For instance, the photodisplacement technique [4] requires modelling of the relationship between the temperature distribution and the resulting displacements [5,6]. Similar problems arise with photoreflectance techniques and more particularly, with photoacoustic methods.

2. Photothermal Radiometry of Induced Carrier Properties

When a semiconductor is excited by a light source with photon energy greater than the bandgap, E_g, carriers are excited. Electrons excited into the conduction band carry excess energy equal to the difference between the photon energy, $h\nu$, and the bandgap. The carriers fall back to the top of the conduction band on the order of a picosecond. On the timescale of the conventional radiometric instrument this can be considered to be instantaneous. These carriers thus remain in their partially excited state at the top of the conduction band, for a characteristic time called the carrier lifetime, τ. The carriers subsequently diffuse through the lattice and recombine releasing energy E_g as heat (assuming the recombination mechanisms are non-radiative). Photoexcitation thus produces both heating and carrier generation in the lattice. The photothermal response in semiconductors is thus considerably more complicated than the situation that applies in non-semiconductors. To summarize, there are three basic processes: (1) Excitation of the carriers followed by instantaneous thermalization with the lattice; (2) Diffusion of the excess carriers; (3) Recombination of electron-hole pairs after a characteristic time, τ.

From the point of view of radiometry processes (1) and (3) lead to heating of the lattice which increases the black body radiation emitted from the sample. Process (2) is another mechanism of energy transport carrying heat away from the site of optical absorption. The optically induced carriers provide an additional source of black body radiation which can be used to measure the electronic properties of a sample - see section IV.

It is shown experimentally in section II that the change in transmission coefficient due to free carrier generation can greatly exceed the radiation arising from temperature changes. For intrinsic semiconductors free carrier generation is the dominant signal generating mechanism, which means that the signal can be predicted from the model of carrier diffusion alone which greatly simplifies quantitative interpretation. A further aid to reliable and simple quantification is that when the one dimensional model for carrier diffusion (see section III.1) is valid for large detector areas the signal change in the transmission coefficient depends on the total number of injected carriers rather than their position in the sample, which means that the signal level is only very weakly affected by changes in carrier diffusion coefficient. When, however, the detector/source dimensions are comparable with the thermal diffusion length it is often difficult to distinguish between signal changes due to recombination or diffusion out of the field of view of the detector.

The straightforward interpretation of radiometric results contrasts with other techniques such as photoreflectance [7] and photodisplacment [4] where both thermal and carrier effects are of similar orders of magnitude. The proviso of large source/detector dimensions is, however, important since only under these conditions does the carrier diffusion cease to affect the signal from the detector.

To summarize we have the rather fortuitous situation that quantitative interpretation on both semiconductors and non-semiconductors is simplified by the absence of conflicting contrast mechanisms. The technique is thus particularly direct and capable of quantitative interpretation as will be demonstrated later, particularly in section IV.

3. Some Applications of Photothermal Radiometry for Semiconductor Analysis

The quantitative application of radiometry will be discussed in section IV. This section aims to give a very brief overview of some of the applications where the technique has been applied. As mentioned in the previous section photothermal radiometry can be used to study any mechanism that alters the carrier concentration, such as surface recombination or lifetime. In fact, the emission qualities can be directly dependent upon the silicon doping density, allowing emissivity measurements to determine carrier concentrations in the region of 10^{18} cm^{-3}. In this work [8] the silicon wafer was uniformly heated and the thermal radiation modulated with a mechanical chopper, providing about 20μm spatial resolution.

For semiconductors photothermal radiometry is able to generate similar information to other thermal wave techniques, but is in addition capable of a more accurate and simpler quantitative interpretation. Some quantitative applications of the technique are demonstrated in section IV, but before concluding this introduction it is worth summarizing some problems to which radiometry has been applied.

One of the earliest applications of photothermal radiometry to semiconductor characterization is the detection of ion implantation damage [9]. The residual damage in the lattice reduces the carrier lifetime and thus reduces the signal level from the sample. This technique has given similar results to photoreflectance and has been compared directly with photodisplacement microscopy. In both these techniques the signal level increases with increasing damage (decreasing lifetime). This is due to a reduction in energy transport from the heat source by the photoexcited carriers, as a direct consequence of reduced lifetime. The heat density, responsible for the photoreflectance and photodisplacement signals is therefore increased. In comparison, the advantage of PTR, however, is that the results are much easier to quantify.

PTR has also been used to examine the surface quality of GaAs wafers following reactive ion etching [9] which showed that etch damage corresponds to a reduced signal (see section IV). The technique has subsequently been used to study reactive ion etching in InP, Si and GaAs [10]. Work using PTR on GaAs was carried out by Mikoshiba and Tsubouchi [11] who have used radiometric imaging in combination with photoluminescence to image wafers. In that work they demonstrated that increased signal level was indicative of non-radiative recombination arising at defects, whereas the PL is sensitive to radiative recombination only. These latter results demonstrate the powerful diagnostic capabilities of PTR, however, care needs to be exercised in determining whether the thermal or carrier mechanism is dominant. If this cannot be established, it is difficult to make a reliable interpretation of the data. Mikoshiba and co-workers have also demonstrated the sensitivity of the instrument to dislocation damage in GaAs and that this sensitivity was enhanced by varying the photon energy just below the bandgap. Peak sensitivity to dislocation density was found for pump wavelengths close to 900nm (bandgap in GaAs corresponds to 867nm).

Quantitative applications of the technique to diagnostics in silicon will be discussed in section IV. Much of this work involves sweeping the modulation frequency of the pump beam in order to obtain a measurement of carrier lifetime. The alternative is to use a short pulsed laser source as used by Cho and Davis [12] where the decay of the radiometric signal was also correlated to carrier density. It should be emphasized, of course, that the essential physics behind pulsed excitation and continuous wave excitation is the same provided the carrier densities do not become so large that Auger recombination starts to dominate.

II. PTR SIGNAL GENERATION

In most cases Photothermal Radiometry detects the excess black body radiation emitted from a sample under thermal excitation and has found wide application in many areas of non-destructive evaluation. When applied to semiconducting samples it has been noted by some workers that the signal response cannot be explained in terms of the commonly used thermal wave physics and some additional signal mechanisms must be brought into play. This new contribution must, of course, produce a change in black body emission from the sample, the dynamics of which have been shown to be typically faster than those attributed to thermal waves. In this section we discuss the signal contribution due to the photo-induced carriers which can be used to explain some of the results presented in section IV.

Ideally a model including both sources of black body emission is desired, the starting point of which is the solution to the coupled diffusion equations. It is demonstrated, however, that in many cases one of the emission mechanisms is dominant and the analysis can be simplified. Furthermore, although the excess carrier distribution acts as a heat source, the excess temperature makes negligible contribution to the excess carrier distribution for most common semiconductors. This enables some decoupling of the thermal equations as shown in the next section.

In addition, a convenient method for obtaining a three dimensional analysis starting from the usual one dimensional solution is presented enabling the effects of three dimensional diffusion to be discussed.

1. Generation of Photoexcited Carrier Waves

A very simple model for electron-hole pair creation, diffusion and decay is used in this analysis to allow particular cases of radiometric signal to be mathematically determined. The model is therefore only satisfied under low injection conditions as it makes no attempt to include high density carrier effects such as Auger recombination. The latter recombination can be included by replacing τ with a new parameter which is a non-linear function of ΔN. The simple one dimensional treatment finds solutions to the following carrier transport equation:

$$D\frac{\partial^2 \Delta N(z)}{\partial z^2} - \frac{\Delta N(z)}{\tau} + G = \frac{\partial \Delta N(z)}{\partial t} \qquad (2)$$

where ΔN is the injected carrier density (m^{-3}), z is the vertical distance from the semiconductor surface (m), D is the ambipolar diffusion coefficient (m^2/s) and τ is the carrier lifetime. The carrier generation term, G, is assumed to originate from the absorption of laser light with sufficiently energetic photons of energy $h\nu$, ie.

$$G = \frac{\alpha \eta I_o \exp(-\alpha z)}{h\nu} \exp(j\omega t) \tag{3}$$

where we have assumed that the light source is intensity modulated at frequency ω, uniformly illuminates the sample and has absorption coefficient α (m^{-1}). The carrier generation efficiency and surface reflectivity can be accounted for by the constant η. In practice, both electrons and holes are generated with equal rates; but for a doped semiconductor one can assume that the injected minority carriers are of most importance, with τ being the minority carrier lifetime. Furthermore, the electronic parameters can themselves be a complex function of doping density and may reflect high density kinetics. Nevertheless, for a semiconductor slab of thickness d, the appropriate boundary conditions at the front and back surfaces are:

$$D \frac{\partial \Delta N(z)}{\partial z} = s_1 \Delta N(0) \qquad \text{at } z = 0 \tag{4}$$

$$D \frac{\partial \Delta N(z)}{\partial z} = - s_2 \Delta N(d) \qquad \text{at } z = d \tag{5}$$

where s_1 and s_2 are the front and back surface recombination velocities respectively (ms^{-1}). [Note that here we have also neglected the time effects normally associated with recombination states. This is a valid approximation if the laser modulation period is shorter than the trap evolution time. For a more detailed consideration see Ref. [7] for example.] The general a.c. solution is found to be:

$$\Delta N(z) = \frac{\alpha \eta I_o}{h\nu D (\alpha^2 - \sigma_e^2)} \left\{ \frac{\gamma_1 \Gamma_2 - \Gamma_1 \gamma_2 \exp[-d(\alpha + \sigma_e)]}{[\Gamma_2 - \Gamma_1 \exp(-2\sigma_e d)]} \exp(-\sigma_e z) \right.$$

$$\left. + \frac{\gamma_1 - \gamma_2 \exp[-d(\alpha - \sigma_e)]}{[\Gamma_2 - \Gamma_1 \exp(-2\sigma_e d)]} \exp[\sigma_e(z - 2d)] - \exp(-\alpha z) \right\} \tag{6}$$

where

$$\Gamma_1 = \frac{D\sigma_e - s_1}{D\sigma_e + s_1} \qquad \Gamma_2 = \frac{D\sigma_e + s_2}{D\sigma_e - s_2}$$

$$\gamma_1 = \frac{D\alpha + s_1}{D\sigma_e + s_1} \qquad \gamma_2 = \frac{D\alpha - s_2}{D\sigma_e - s_2}$$

These coefficients determine the effects of the front and back surfaces and in many ways can be likened to reflection coefficients at the material interfaces. The term σ_e is the complex propagation coefficient associated with the photoexcited carriers and is given by:

$$\sigma_e = \left[\frac{(1 + j\omega\tau)}{D\tau}\right]^{\frac{1}{2}} \qquad (7)$$

The form of the carrier propagation coefficient tells us that the photoexcited carrier plasma behaves like a critically damped wave for modulation frequencies where $\omega\tau > 1$. [Physically, for the higher frequencies one finds that the photoexcited carrier density response time is longer than the laser modulation period, and so changes in modulation frequency affect the a.c. component of the carrier density.] For $\omega\tau \ll 1$ the propagation coefficient is independent of modulation frequency. More intuitive information can be derived from Eq. (6) if we approximate the solution to a special case. For example, if the laser absorption coefficient, α, and the sample thickness, d, are assumed to tend to infinity, and the effect of front surface recombination is ignored, then $\Delta N(z)$ reduces to:

$$\Delta N(z) = \frac{\eta I_o}{h\nu D\sigma_e} \exp(-\sigma_e z) \qquad (8)$$

Mathematically, the form of Eq. (8) is similar to that normally derived for the propagation of thermal waves. The important difference arises from the frequency dependence of the carrier propagation coefficient. In the limit $\omega\tau \gg 1$ the carrier wave behaves like its thermal counterpart. It has a diffusion length given by $(2D/\omega)^{1/2}$ and a peak density which varies as $\omega^{-1/2}$. Furthermore, if the signal arises predominantly from the photoexcited carriers, it may be possible to perform depth profiling by varying the carrier propagation length and obtain information from different layers in the semiconductor. In the other limit, $\omega\tau \ll 1$, the carrier penetration and peak density remains unaffected by changes in modulation frequency. The transition between these two limits can be used to determine carrier lifetime data as discussed in section IV.1.

2. Generation of Thermal Waves

Intensity-modulated light incident on the semiconductor will generate both thermal and photoexcited carrier waves. The thermal wave solution can be derived from the following coupled equation:

$$\frac{\partial^2 \Delta T(z)}{\partial z^2} + \frac{\eta(h\nu - E_g)}{h\nu K}\alpha I_o \exp(-\alpha z) + \frac{E_g \Delta N(z)}{K\tau} = \frac{j\omega \Delta T(z)}{\beta} \quad (9)$$

were K is the thermal conductivity of the semiconductor slab (W/K), β is the thermal diffusivity (m^2/s) and $\Delta T(z)$ is the resulting a.c. temperature distribution. The two heat source terms arise from:
(a) Carrier relaxation to the conduction band edge. In this case the carriers are assumed to be excited to a level well above the band edge. After a short time (~ psec) they give up an energy $(h\nu - E_g)$ to the lattice as heat. Due to the short relaxation time carrier diffusion effects can be ignored.
(b) Carrier recombination. After a longer period τ the carriers give up an energy E_g to the lattice in the form of heat. Carrier diffusion is accounted for in this term via $\Delta N(z)$.

There is another heat source term which arrives from the boundary conditions of flux continuity. For the slab semiconductor, assuming no thermal conduction to surroundings, the boundary conditions are:

$$-K\frac{\partial \Delta T(0)}{\partial z} = s_1 \Delta N(0) E_g \quad : \quad \text{Front surface}$$

$$K\frac{\partial \Delta T(d)}{\partial z} = s_2 \Delta N(d) E_g \quad : \quad \text{Back surface}$$

A full solution can be obtained by solving Eqs. (6) and (9) simultaneously, and extensions to include the effect of additional semiconducting layers can be included by increasing the number of equations and modifying the boundary conditions appropriately. For simplicity we shall consider the solution when the laser absorption coefficient and the sample thickness tend to infinity, ie.

$$\Delta T(z) =$$

$$\left\{ \frac{\eta s_1 E_g I_o}{h\nu K \sigma_t (D\sigma_e + s_1)} + \frac{\eta \sigma_e E_g I_o}{h\nu K \sigma_t \tau (D\sigma_e + s_1)(\sigma_e^2 - \sigma_t^2)} + \frac{\eta(h\nu - E_g)I_o}{h\nu K \sigma_t} \right\} \exp(-\sigma_t z)$$

$$-\frac{\eta E_g I_o \exp(-\sigma_e z)}{h\nu K\tau (D\sigma_e + s_1)(\sigma_e^2 - \sigma_t^2)} \qquad (10)$$

with $\sigma_t = (1+j)/\mu$ where μ is the thermal diffusion length given by $(2\beta/\omega)^{1/2}$. The terms in the brackets represent the contribution of thermal waves generated by the three different heating mechanisms, ie. surface recombination, bulk recombination and carrier relaxation respectively. The remaining term arises from carrier diffusion and implies a modification to thermal wave propagation. For the additional case of $E_g \approx h\nu$ and $\tau \to 0$ we note that:

$$\Delta T(z) = \frac{\eta I_o}{K\sigma_e} \exp(-\sigma_t z) \qquad (11)$$

which, as one might expect, is exactly equivalent to thermal wave propagation in a non-semiconductor.

3. Black Body Emission Mechanisms

All objects exist in a continual state of radiation transfer. An object in thermal equilibrium with its surroundings will have a photon emission rate per wavelength interval exactly equal to the rate of photon absorption - this is a statement of Kirchoff's Law. At room temperature, Planck's law predicts a radiation exchange in the infrared, predominantly between 8 - 14 μm wavelength (called black body radiation). A focused laser beam incident on a sample will locally disturb this thermal equilibrium, causing an excess thermal energy normally dissipated by heat conduction into the bulk of the material away from the point of laser absorption. However, the thermal imbalance is also accompanied by an increase in the radiation exchange from the object into its surroundings. Intensity modulation of the incident laser power therefore produces an a.c. modulation in the black body emission from the point of laser excitation - this is the signal source in photothermal radiometry. The a.c. radiated power is found from the Stefan-Boltzmann law to be: $\Delta q = 4\epsilon\sigma T^3 \Delta T$ (watts/cm^2), where ΔT is the a.c. temperature fluctuation responsible for the a.c. emission. A typical example of black body radiation arising from a thermal wave is shown in Fig. 1.

Returning to Kirchoff's law, a change in the black body emission from an object could arise from a change in the amount of radiation it absorbs. Therefore an increase in an object's absorption coefficient must lead to an increase in black body emission to remain at thermal equilibrium. There is in fact a direct relation between absorptivity and emissivity expressed by

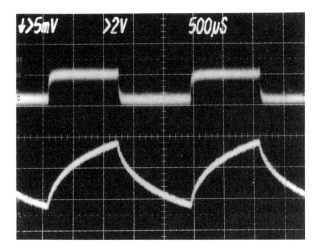

Fig. 1 Modulated laser power reflected from a sample of rubber (top trace) and resulting change in sample temperature via black body emission (lower trace).

Kirchoff's laws. For semiconductors we must first distinguish between two distinct electronic transitions which are responsible for absorption; band-to-band transitions and intra-band transitions. Generally the production of photoexcited carriers relies upon transitions between valence and conduction bands as laser photons are absorbed. However, the absorption of infrared radiation within the semiconductor is much weaker and relies upon lower energy transitions within the same energy band (intra-band). Therefore increasing the density of free carriers by photoexcitation makes more carriers available for intra-band transitions and so increases the infrared absorption. Assuming we have not moved far from equilibrium one would also expect a corresponding increase in black body emission. This change in emission, arising from free carrier injection was demonstrated by Ulmer and Frank [13] using a p-n junction. These authors concluded that the basic principles of Kirchoff's law were still applicable under conditions of strong carrier injection. In this section we will, therefore, derive an expression for the changes in black body emission from a semiconductor using the free carrier absorption coefficient as a starting point. Indeed, one can see from the outset that the excess charge carriers oscillate with random thermal motion and thus act as sources of radiation. It is, however, more useful to consider the effect of excess carriers in terms of Kirchoff's Radiation Law of detailed energy balance which states that the emission of radiation at a given wavelength and temperature is exactly equal to the absorption at the same wavelength and temperature. This law arises by noting that only in this way can a dynamic thermal equilibrium between an object and the background be achieved.

To make our discussion more definite we shall refer to the semiconductor sample of thickness, d, shown in Fig. 2. As discussed in Ref.

[14], the sample may be considered as a partially transparent body with radiation generated and absorbed within the sample. The front surface has an internal infrared reflectivity R. We will also consider the back surface of the semiconductor to scatter incident radiation so that radiation reflected into the semiconductor from the front face is assumed lost. This assumption is likely to be valid for wafers where the back surface is not polished and the area of the infrared detector is small. The back surface and the backing material can nevertheless have a very significant effect on the magnitude of the photothermal signal when the back surface is smooth [15], as discussed in section IV.4.

From Kirchoff's Radiation Law, which states that the radiation emitted from each region is equal to the absorption, we note that the radiation emitted from the thin radiating slice in Fig. 2 will be given by:

$$\alpha_{ir}(\lambda,z) \, Q_s(\lambda,\theta) \, \delta z$$

where $\alpha_{ir}(\lambda,z)$ is the absorption coefficient of the detected infrared radiation as a function of wavelength and depth. [It should be noted, of course, that the absorption coefficient of the laser *pump* beam is only relevant in determining the temperature and carrier distribution, which then determine the precise form and variation of $\alpha_{ir}(\lambda,z)$.] The radiation density, $Q_s(\lambda,\theta)$ is a function of wavelength and sample temperature, θ, and may for ideal samples be determined by Planck's radiation law.

As the emission propagates through the semiconductor it is attenuated in accordance with Lambert's Law. Applying Kirchoff's radiation law to every point within the semiconductor gives the total radiation leaving the front surface. For the one dimensional case:

$$\int_{\lambda_1}^{\lambda_2} \int_0^d (1-R) \, \alpha_{ir} \exp(-\alpha_{ir} z) \, Q_s(\lambda,\theta) \, dz \, d\lambda = \int_{\lambda_1}^{\lambda_2} (1-R) \, [1-T_r] \, Q_s(\lambda,\theta) \, d\lambda \, . \tag{13}$$

Fig. 2 Schematic one dimensional diagram showing emission and attenuation of radiation in a semi-transparent sample.

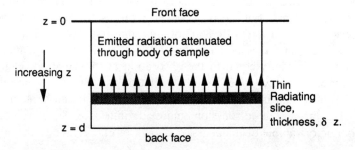

T_r is the infrared transmission coefficient = $\exp(-\alpha_{ir}d)$; λ_1 and λ_2 define the spectral limits of the infrared detector.

When the laser illuminates the sample the transmission coefficient will be modified to some value T_r^*, and the increase in infrared emission from the object will be:

$$\Delta q = \int_{\lambda_1}^{\lambda_2} (1-R) \, [T_r - T_r^*] \, Q_s(\lambda,\theta) \, d\lambda \tag{14}$$

assuming that any variation in reflectivity, R and the sample temperature can be ignored at this stage. Equation (14) gives the a.c. radiated power due to carrier injection. The modified transmission coefficient T_r^* can be found from the change in infrared absorption due to the excess carrier distribution. Thus the absorption is likely to decrease into the sample (z increasing) as the injected carrier density decreases. In our simplified one dimensional treatment we approximate:

$$T_r^* = \exp\left[-\int_0^d \alpha_{fc}(z) \, dz - \alpha_{ir}d\right] \tag{15}$$

where α_{ir} is the unperturbed absorption coefficient and $\alpha_{fc}(z)$ represents the increase in absorption due to the photoexcited carriers. From a classical model for wave propagation in a free carrier plasma we have:

$$\alpha_{fc}(z) = \frac{\Delta N(z)\lambda^2 \, e^3}{4\pi^2 c^3 n\varepsilon_o m^{*2}\mu} = \xi \Delta N(z) \tag{16}$$

which relates to the injected carrier density, $\Delta N(z)$, derived in section II.1. For weak intrinsic absorption (ie. lightly doped samples ~ 10Ωcm) and low injection density we may expect $T_r \approx 1$ and, to a first order approximation we can write

$$\Delta q \propto \int_0^d \Delta N(z) \, dz \, . \tag{17}$$

suggesting that the emission depends upon the total number of carriers injected. However, for strong absorption or high injection the emission will depend upon the carriers lying closer to the sample surface.

The radiometric signal derived from the infrared detector originates from both thermal wave and injected carrier emission. In many respects, the application of the technique to semiconductors depends upon which effect provides the dominant contribution. Our observations show that for lightly doped samples of Si and GaAs the injected carrier contribution dramatically dominates the thermal contribution. [In many other cases it may be possible to find a laser modulation frequency which allows imaging via injected carriers due to the different frequency dependence of the two components.] This is demonstrated by the oscilloscope trace in Fig. 3, which shows the black body emission arising from a silicon sample using PTR. The response time is much faster than that which can be attributed to thermal wave effects similar to those in Fig. 1, and we conclude that the response is due to the more rapid rates of carrier generation and decay. Another simple interpretation concludes that the laser is able to modulate the sample emissivity in a response time which is dependent upon the injected carrier lifetime. Furthermore, as is clear from Eq. (14), it is possible to enhance the magnitude of the a.c. emission by increasing the sample temperature. Unduly large changes in temperature may be expected to affect quantitative measurements of material properties. The fact that the signal can be increased by raising the temperature only 20°C above ambient is useful in cases where the signal-to-noise ratio is poor.

Fig. 3 Modulated laser power reflected from surface of silicon sample (top trace) and resulting change in infrared emission (lower trace).

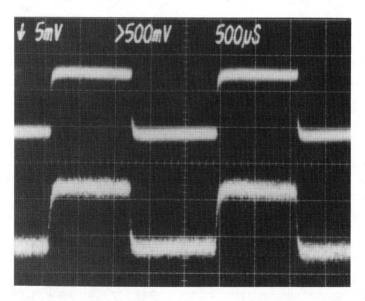

4. Three Dimensional Extension

In this section a simple and convenient method of obtaining three dimensional solutions to carrier/thermal wave propagation is demonstrated. For convenience, equations (2) and (9) will be written in the generalized form, relating to either carrier density or temperature.

$$\nabla^2 \phi(x,y,z) + \alpha \, \psi(x,y) \, e^{-\alpha z} = \sigma^2 \phi(x,y,z) \tag{18}$$

where σ is the complex propagation coefficient and $\psi(x,y)$ represents the distributed source term with absorption coefficient α. Taking the two dimensional Fourier Transform gives:

$$\frac{d^2 \Phi(f_x,f_y,z)}{dz^2} + \alpha \, \Psi(f_x,f_y) \, e^{-\alpha z} = k^2 \, \Phi(f_x,f_y,z) \tag{19}$$

where $\Phi(f_x,f_y,z)$, $\Psi(f_x,f_y)$ are the transformed variables and k is the Fourier space propagation coefficient equal to $[\sigma^2 + 4\pi^2(f_x^2 + f_y^2)]^{1/2}$. The general solution to the above equation with transformed boundary conditions is equivalent to the normal one-dimensional solution obtained from either Eq. (2) or (9), but with σ replaced by k. For example the standard one-dimensional solution for an infinite solid is:

$$\phi(z) = \frac{\alpha I_o}{(\alpha^2 - \sigma^2)} \left\{ \frac{\alpha}{\sigma} e^{-\sigma z} - e^{-\alpha z} \right\} \tag{20}$$

with I_o being the optical power density. The three-dimensional solution is conveniently obtained from the inverse Fourier transform of:

$$\Phi(f_x,f_y,z) = \frac{\alpha \Psi(f_x,f_y)}{(\alpha^2 - k^2)} \left\{ \frac{\alpha}{k} e^{-kz} - e^{-\alpha z} \right\}. \tag{21}$$

In practice it is necessary to use a focused infrared detector to collect sufficient black body radiation from the point of laser excitation on the sample. The emitting area seen by the detector is a useful parameter to include into simulations. If one can define a detector function $g(x,y)$, which describes the detector responsivity projected onto the sample surface, then the detected signal response is given by:

$$\int\int_{-\infty}^{\infty} g(x,y) \, \phi(x,y,0) \, dxdy \tag{22}$$

For symmetric functions one can make use of Rayleigh's theorem [16] and convert the integral to the Fourier domain,

$$\int\int_{-\infty}^{\infty} G(f_x,f_y)\Phi(f_x,f_y,0)\, df_x df_y \ . \tag{23}$$

Equation (23) contains the Fourier transforms of the detector spatial responsivity, the source distribution and the relevant one-dimensional solution, and can be conveniently solved using numerical methods. Furthermore, for the generalized solution of $\Phi(f_x,f_y,z)$ discussed above one realizes that there is a reciprocal relationship between source and detector distributions, implying that one dimensional behavior can be achieved using either a sufficiently broad laser source or large area detector. A similar three-dimensional analysis is used in section IV.1 where the implications of lateral diffusion are discussed in more detail.

III. THE PTR MICROSCOPE

1. 1-D Versus 3-D Arrangement

The spatial resolution is of particular significance when designing a PTR instrument. Small structures in the sample often give localized contrast, so that in some cases the smallest detectable signal is determined by the source size. However, it will be shown in section IV.1 that the effect of sideways diffusion on photoexcited carriers can strongly influence the characteristics of the PTR response, and in general make the direct measurement of electronic properties difficult to obtain with sufficient confidence. For accurate determination of the electronic properties it may therefore be desirable to work with an unfocused pump beam several carrier diffusion lengths across. The one-dimensional model for the PTR response can then be used to extract physical data from experimental results with greater confidence. This highlights the direct conflict between high spatial resolution and accurate measurement of material properties. In our work we have therefore constructed two PTR instruments; one best suited for microscopic analysis (focused spot diameter ~ 2µm) and another for measurement of electronic properties (spot diameter > 1mm). The higher resolution instrument, which will be described as the PTR microscope is shown in Fig. 4 and is discussed in the following section. Its use has been predominantly restricted to microscopic examination of semiconductors for

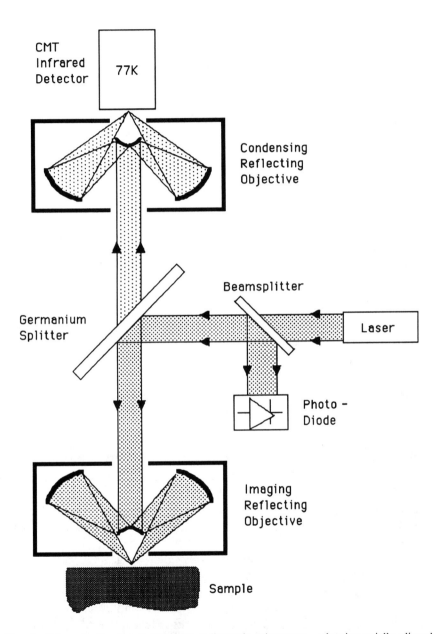

Fig. 4 Schematic for the photothermal radiometric microscope - showing axially aligned reflecting objectives and germanium beam splitter.

qualitative assessment and imaging - see section IV. The one-dimensional system has been reserved for the measurement of carrier lifetimes.

2. Configuration and Performance

The pump beam is derived from a laser diode enclosed in a collimating pen (Philips CQL series) providing up to 20mW cw power at 860 nm. Intensity modulation is achieved by direct control of the laser bias current. The laser light is incident at 45° to a germanium beam splitter specially coated to provide >90% reflectivity at 860 nm, while maintaining >85% transmissivity for longer wavelengths between 8-12µm. (The broadband response is desirable to make best use of the i.r. detector spectral bandwidth). The reflected light enters the aperture of the lower reflecting objective (*imaging objective*, N.A. = 0.5, Ealing Beck), which is used to focus the pump beam down to a spot size of ~ 2µm. Back scattered light from the sample is returned to a photodiode to allow scanning optical images of the samples to be taken. Black body radiation from generated thermal waves and photoexcited carriers is collected by the imaging objective, collimated and transmitted by the germanium beam splitter. The second reflecting objective (*condensing objective*, N.A. = 0.5) is used to provide a diffraction limited focus of black body emission onto the infrared detector element. Both objectives are gold-coated to enhance infrared reflectivities, and do not present the usual problems associated with chromatic aberration and absorption in refracting lenses.

The choice of infrared detector is critical to the signal-to-noise ratio performance of the instrument. In our case we have used a photoconductive, 50µm x 50µm Cadmium Mercury Telluride detector with Joule-Thompson cooling [GEC-Marconi]. The detector has an integral germanium window and 60° cold field-of-view restrictor. Its spectral range lies between 8-14µm wavelength and has an NEP = 3×10^{-13} W/(Hz)$^{1/2}$ and 0.4µs response time. After (low noise) amplification the detector output is measured using a dual phase lock-in amplifier, with both amplitude and phase data recorded simultaneously using an IBM PC. Stepping motors are used to scan the sample stage with a minimum step size of 2µm. Scanning is also controlled by the PC enabling a number of point measurements to be taken over a semiconductor wafer or complete point-by-point images to be taken and displayed. The laser modulation frequency can also be set by the PC. To achieve focus the whole of the microscope section can be moved independently of the sample stage in the vertical direction. It is worth noting that deliberate defocusing can be used to effect some degree of one dimensional behavior at the inevitable cost to signal-to-noise ratio. Alternatively a lower N.A. imaging objective can be used sacrificing i.r. collection efficiency. For non-semiconductor samples with emissivity close to unity the minimum detectable temperature change with a 1 Hz electrical bandwidth is estimated as 1.7×10^{-4} K.

IV. SEMICONDUCTOR ASSESSMENT

The successful production of electronic devices relies upon a detailed knowledge of the properties of the process material. Indeed it is often insufficient to guarantee only the quality of the starting material, as many processing stages modify the properties of the semiconductor. Full materials characterization using conventional electrical methods (ie. sheet resistivity, Schottky barrier, DLTS, PITS, etc.) can be both expensive and destructive to the test sample. There is an increasing need, therefore, for assessment techniques that can be used between processing stages, without damaging the process material. PTR is a non-contacting method and could be used as a rapid on-line process monitoring technique for both qualitative and quantitative measurement.

In section II.3 it was shown that for lightly doped silicon samples the PTR signal arises predominantly from the photo-induced carriers rather than from generated thermal waves. In this situation one might expect that the technique is able to provide information about the electronic properties of the sample instead of its structural/thermal characteristics. In this section we aim to demonstrate that in many cases information provided by the technique can be interpreted as either being due to electronic or thermal features, since one mechanism is likely to strongly dominate. Results from sections 1-4 below deal with cases where the carrier contribution is dominant, whereas sections 5-6 demonstrate cases where the thermal wave emission dominates. In these last two sections the results can be interpreted using the classical thermal wave arguments discussed elsewhere in this text.

1. Direct Determination of Carrier Lifetime

A valuable guide to the quality/purity of semiconducting material is the minority carrier lifetime. In this section it is demonstrated that for lightly doped silicon samples it is possible to determine directly the carrier lifetime of the injected carriers. The method adopted operates in the frequency domain, although a pulsed approach could also be used. However, the data interpretation may be more convenient in the frequency domain. The response time of the infrared detector (~ 0.4µs in our case) sets the lower limit on lifetime measurement. The method is therefore unsuitable for GaAs where the lifetimes are typically much shorter. In order to understand how the method works we need to look back at Eqs. (6) and (17). For simplification we shall consider a semi-infinite sample ($d \rightarrow \infty$), with infinite absorption coefficient for the pump beam. In this case we have:

$$\Delta q = \frac{\eta I_o}{h v \sigma_e (D\sigma_e + s_1)} \quad (23)$$

Taking logarithms we obtain:

$$\log_{10} |\Delta q(\omega)| \propto A - \log_{10}\left(\frac{1 + j\omega\tau}{\tau}\right)\left[1 + \frac{s_1 \tau}{L_e(1 + j\omega\tau)^{1/2}}\right], \quad (24)$$

where A is a constant which is normalized to the experimental data and L_e is the injected carrier diffusion length equal to $(D\tau)^{1/2}$. Initially we shall ignore the 2nd term in the square brackets. For low frequencies where $\omega\tau \ll 1$ the PTR signal remains constant and equal to $A+\log_{10}(\tau)$. For $\omega\tau \gg 1$ the signal falls for increasing $\log_{10}(\omega)$ with a constant slope of -1. If the second term dominates, we have a signal which is proportional to the surface recombination velocity at low frequencies and a slope of -1/2 for $\omega\tau \gg$. It is possible therefore to obtain τ, and in some cases s_1 from the frequency characteristics of the sample.

Our measurement methodology is then as follows. Using the one dimensional PTR instrument we record $\log_{10}$. (amplitude) over the frequency range 100 Hz to 100 kHz. The computer simulation takes this experimental data and compares it with the theoretical model described in section II, performing a least square error analysis to obtain estimates of τ and s_1 for other fixed parameters. For the frequency range chosen it is possible to determine reliable estimates of τ within the range 5 -500 μs. From Equation (24) we note that the signal response will be sensitive to surface recombination effects at low frequencies when $s_1 > L_e/\tau$, ie. when the carrier diffusion length is small so that the carriers remain close to the sample surface, or when the lifetime is sufficiently long to allow surface recombination to dominate. For this condition s_1 must be typically greater than 10 m/s for silicon. After the break point a slope of -1/2 will be obtained if $s_1 \gg (\omega D)^{1/2}$. For frequencies up to 100 kHz this corresponds to $s_1 \gg$ 50m/s. Before leaving Eq. (24) it is worth remembering that these simple conclusions are only valid for the one-dimensional case and for strong absorption of the pump beam.

The test sample consists of a long lifetime silicon wafer sample (250 Ωcm, 75mm dia.) produced for power transistor fabrication[1]. The wafer was shadow masked using a molybdenum slug and electron-beam irradiated

---

[1] It should be mentioned that radiometry is particularly suitable for examination of materials for power devices because the infrared radiation is transmitted through the whole body of the wafer, which in these devices is integral to the functioning of the device. In lower power VLSI applications one is generally only concerned with the quality in the top few microns.

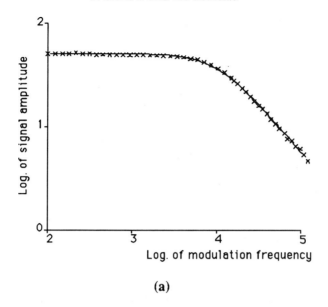

(a)

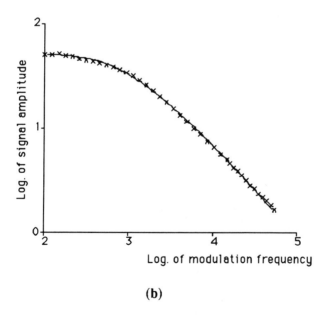

(b)

Fig. 5 a) Experimental (xxx) and theoretical simulation (—) results for short lifetime silicon region, one-dimensional instrument. Parameter estimates are $\tau = 17.7\mu s$ and $s_l = 4.1$ m/s. b) Experimental (xxx) and theoretical simulation (—) results for long lifetime silicon region, one-dimensional instrument. Parameter estimates are $\tau = 443\mu s$ and $s_l = 8.3$ m/s.

to reduce the lifetime over the exposed regions of the wafer [17]. Experimental data was taken at various points on the sample using both the one dimensional and three-dimensional PTR systems. Using the simulations we obtain the curves presented in Figs. 5 and 6.

In Fig. 5a we show a typical result taken with the one-dimensional system for the irradiated region of the sample. Both experimental data (x's) and theoretical data (line) are superimposed. The corresponding simulation parameters were: laser beam width = 5.0mm, detector size = 1.0mm^2, and $\alpha = 3.0 \times 10^6$ cm^{-1}. The simulation produces an excellent fit to the experimental data for estimates of $\tau = 17.7\mu s$, and $s_l = 4.1$m/s. Repeating these measurements over the masked region of the wafer provides the result shown in Fig. 5b, with $\tau = 443\mu s$ and $s_l = 8.3$m/s. As expected the irradiation process dramatically reduces carrier lifetime without significantly modifying the surface recombination velocity, the estimates in the two regions being close to those expected by the manufacturer. The high lifetime, in

Fig. 6 Results from the three-dimensional instrument at same location on sample as in Fig. 5(a). Same values of τ and s_l obtained for instrument parameters: laser beam width of 2μm and effective detector width of 86μm.

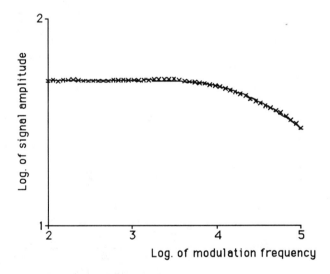

excess of 400μs, of the virgin wafer is a typical requirement of power devices. For the values of s_l suggested by the simulation one would not expect the experimental data to be particularly sensitive to variations in surface recombination velocity, and under these conditions more confidence is placed on the estimate of carrier lifetime data.

Experimental simulation results for the three-dimensional instrument are shown in Fig. 6 for the irradiated section of the wafer at the same location as that for Fig. 5a. The three dimensional simulation produced the same values of $\tau = 17.7\mu s$ and $s_l = 4.1$ m/s for the following parameters defining the optical configuration: gaussian laser beam width = 2μm (laser intensity = 100mW) and detector size = 86x86μm² (fitted parameter). Results for the masked region are not presented here, but the simulation obtained $\tau = 62\mu s$ and $s_l = 1.5$m/s, which are dramatically different from that obtained with the one-dimensional instrument for the same point on the sample.

Firstly we make the observation that the one-dimensional simulations produce curves with two well defined sections as discussed above, whereas the three-dimensional simulations produce curves whose knee extends over a greater frequency range. Furthermore the gradient beyond the knee ($\omega\tau \gg 1$) is reduced. This is a result of the additional effect of carriers diffusing to regions outside the field-of-view the detector. This process is difficult to distinguish from carrier recombination and therefore broadens the region of the curve critically dependent upon lifetime. This is a characteristic of the higher resolution instrument and makes the simulation exercise less reliable, the estimated parameters being highly sensitive to the exact laser beam and detector size defined at the start of the simulation. This is unsatisfactory from the point of view of the measurement accuracy since one requires a technique insensitive to the exact parameters of the instrumentation. Further inaccuracies for this measurement could also be attributed to the neglect of Auger recombination in our model, which clearly plays a more significant role for the tightly focused pump beam and higher lifetime case. However, the simulations in Fig. 5a and Fig. 6 are in good agreement when the carrier diffusion length is shorter.

2. Ion Implantation Study

An important process in the fabrication of the FET family of devices is ion implantation. This processing step allows an exact amount of dopant to be implanted into a semiconductor wafer. The dopant consists of ionized-projectiles having sufficient energy to penetrate the wafer surface to a depth between 100 to 10,000Å. Crucial to maintaining device yield across the whole wafer is the uniformity of the implant dose. The difficulty is to find methods to determine both implant density and uniformity over a wafer in a non-destructive manner. Techniques currently available include Secondary Ion Mass Spectroscopy (SIMS), Rutherford Backscattering Spectroscopy (RBS), four point probe and sheet resistance measurements. However, all of these techniques either require pre-processing or can be classed as destructive. SIMS and RBS are also particularly costly to install and operate. This study shows that PTR may offer some advantages over these methods, being non-destructive, and offering rapid high resolution measurements.

Firstly the effect of lattice damage was investigated. Several silicon wafers were obtained which were ion implanted with silicon ions within the dose range of 10^{10} to 10^{15} ions/cm^2. The use of silicon ions ensures that the material purity remains constant, while the effect of lattice damage can be observed. The wafers were n-type with resistivity 3 - 27 Ω cm and (100) orientation. In all these samples only half the wafer was implanted, so that the unprocessed side could act as a signal reference. An example of the line scans obtained with the three-dimensional PTR instrument is shown in Fig. 7. The modulation frequency was set at 10 kHz, such that $\omega\tau < 1$ and the signal level is assumed to be proportional to τ (see Eqs. 23 and 24). Over the

Fig. 7 Amplitude line scan over a half-implanted silicon wafer, showing reduced signal level due to ion implantation damage. Implant dose = 10^{15} silicon ions/cm^2 at 180 keV.

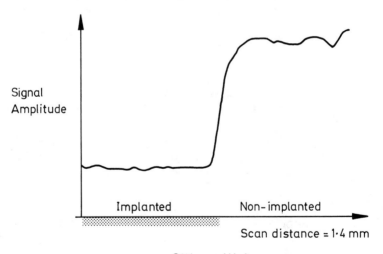

implanted side of the wafer the signal level decreases. This can be explained by assuming that the implantation produces a stratified region of lattice damage below the silicon surface which results in a decrease in injected carrier lifetime. Despite the fact that the damaged region is very thin the implantation has caused a significant reduction in the effective carrier lifetime, simply because τ in this region will be of the order of nanoseconds due to the lattice disorder[2]. This result was repeated using a 488nm beam which showed a very dramatic increase in contrast between implanted and non-implanted sides of the wafer. This enhanced sensitivity is due to the reduction in the laser absorption length going from 860nm to 488nm (ie. ~ 10μm to <1μm) which brings the photoexcitation process closer to the damaged region responsible for the contrast. The choice of pump wavelength therefore determines the relative sensitivity to surface or bulk effects.

The same sample was also examined using photothermal displacement [4]. For this technique the induced carriers have two distinct effects. Firstly they act as a heating source which produces thermal expansion, secondly they modify the refractive index which similarly modifies the interferometer output. For most cases the former effect dominates. This means that the displacement signal obtained for damaged wafers is greater as the lower lifetime ensures that carriers recombine closer to the surface giving up their heat to the lattice - which in turn leads to a larger displacement on the surface. Carriers which diffuse deeper into the material, prior to recombining, give a comparatively small peak displacement on the sample surface. The line scan is thus complementary to that shown in Fig. 7, ie. the implanted side

Fig. 8 PTR signal as a function of implant dose for silicon ions implanted into silicon sample at 180keV - measure of lattice disorder.

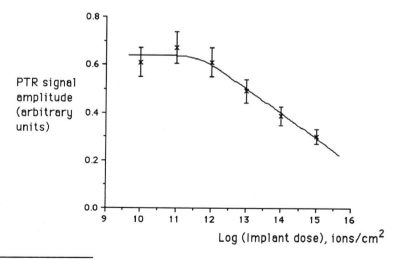

---

[2] If the thermal diffusion length is very small compared to the extent of the implanted region, w, a region of reduced carrier lifetime cannot be distinguished from a region of reduced lifetime. In fact, the excess recombination velocity is given simply by w/τ.

produced a higher signal level. The degree of contrast between either side of the implanted sample was similar to that obtained using PTR. Fig. 8 shows how the PTR signal level varies with implant dose. Note that the technique is more sensitive to dose levels above 10^{13} ions/cm^2 (these results were taken at 860nm).

In a second study [18] the effect of arsenic implantation was considered for both n- and p-type silicon wafers [(100) orientation, 2-4 Ω cm]. The As ions were implanted at 400keV for doses between 10^{12} -10^{15} ions/cm^2. The estimated projected range of the implant is 0.22μm with a standard deviation of 0.078μm. The samples were examined before and after furnace annealing. The annealing was performed under a nitrogen atmosphere for 10 minutes. The furnace temperature was varied between 600°C and 1000°C.

Fig. 9a shows how the PTR signal varies against dose for the as-implanted, non-annealed wafers. As expected from the first set of results the PTR amplitude decreases with increasing lattice disorder, being more sensitive to the higher dose range. The effect of furnace annealing is shown in Fig. 9b. In this case the results are quite different in that the predominant contrast mechanism for the annealed samples is not the recombination dynamics of the laser photoexcited carriers. What distinguishes the two cases is that the relative level of carrier injection has changed. For the un-annealed samples, the laser injected carrier density (approximately 10^{17} cm^{-3}) exceeds that of the background concentration (high-level injection); for the annealed samples, the injected carrier density is small compared with the donor concentration and thus the curves in Fig. 9b reflect the implanted donor activation kinetics (as well as anneal damage). This is most clearly demonstrated for the n-type wafer series where the curve reflects the increased activated implant majority carrier concentration and its effect on emitted black body radiation via the free carrier emission mechanism, ie. an increased emissivity. The curve obtained for the arsenic implanted p-type wafers is more complicated to explain since the donor implantation creates a p-n junction beneath the wafer surface. Photo-induced carriers produced by the laser pump beam are likely to be swept across the depletion regions of this junction. Furthermore, as the wafer has low resistivity the p-side depletion region will extend several microns into the substrate. The black body emission is therefore likely to be reduced due to the fact that injected carriers will be accelerated deep into the substrate and therefore the signal response becomes more dependent upon the junction dynamics.

Other results have shown that the technique is sensitive to ion implant dose variations in GaAs wafers. Similarly the conclusion from these results points to the contrast arising predominantly from the extent of lattice damage rather than ion species.

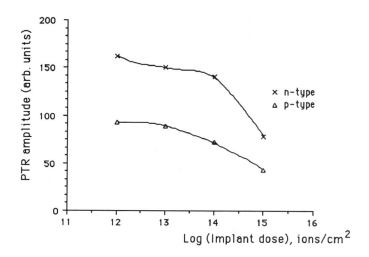

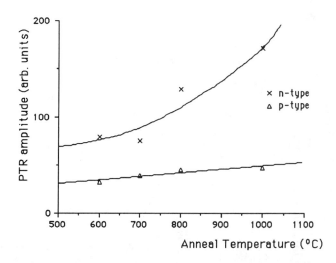

Fig. 9 a) PTR signal amplitude as a function of arsenic implant dose in silicon (ion implant energy 400keV). b) PTR signal amplitude as a function of sample anneal temperature. Implantation parameters before anneal were ion dose 1×10^{15} atoms/cm^2, implant energy 400keV, and implant species arsenic.

3. Effect of Surface Recombination

In this section two examples of PTR measurements on GaAs are presented which demonstrate the sensitivity of the technique to changes in surface recombination. It is believed that similar measurements could be used to determine surface quality. The first results relate to the effect of energetic ion bombardment during reactive ion etching (RIE) of semiconductor samples. The second example shows the effect of improved surface recombination due to the presence of a GaAlAs capping layer on a GaAs substrate. As shown in section IV.1, via Eq. (24), the PTR signal response will be proportional to the surface recombination velocity when $s_l > L_e \tau$, which is the case which normally exists in GaAs (and heavily doped silicon) due to the comparatively short carrier lifetimes.

3.1 Reactive Ion Etching

Crean et al. [10] conducted an experimental study of defects induced in silicon (n-type, 2-4 Ω cm, (100) orientation) and gallium arsenide (n-type, (100) orientation) as a function of isochronal bias voltage, results of which were correlated with Raman spectroscopy measurements. In that work the wafers were wet etched prior to loading into the etch chamber to ensure an undamaged and contaminant free surface. The silicon wafers were subjected to isochronal (20 min.) etching in an argon plasma at a chamber pressure of 15mTorr using a PLASMA TECHNOLOGY reactive ion etch system. The GaAs wafers were similarly etched but in a separate RIBE reactor. The operating pressure was 5 x 10^{-4} mbar.

Fig. 10 Experimental PTR amplitude variation as a function of applied dry etch voltage for 2 x 10^{18} cm^{-3}, n-type GaAs substrates. (From Ref. [10], with permission).

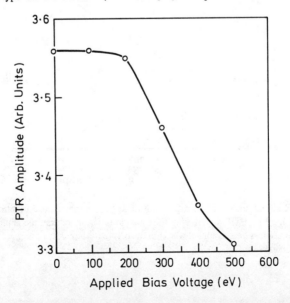

Figure 10 shows the results of the GaAs isochronal etch as a function of applied bias voltage. The general trend demonstrates a decrease in signal with increasing bias. This is consistent with the assumption that increasing bias tends to extend the range and severity of lattice damage in the near surface of the substrate. It is interesting to note however that there appears to be a threshold for induced damage between 200 - 300 eV, which may be attributed to the formation of electron traps. Similar results were obtained for silicon substrates which showed a more monotonic decrease in signal with bias voltage.

Other studies, using the inert gas ions: argon, neon and helium, have demonstrated that the technique is able to measure an increase in surface damage due to increasing ion mass [9].

3.2 GaAs/GaAlAs Interface

In this section an interface between GaAs and a GaAlAs layer grown by Chemical Vapor Deposition (CVD) is examined. The reason for presenting this result is to demonstrate that PTR could provide a useful tool for assessing the quality of grown interfaces via its sensitivity to recombination velocity. The test sample structure and the corresponding PTR line scan data is shown in Fig. 11.

On the exposed GaAs side of the sample the optically injected carriers recombine heavily at the GaAs/air interface, which has a surface

Fig. 11 The GaAs/GaAlAs sample structure, together with the PTR line scan - showing improvement in carrier recombination at the GaAs/GaAlAs interface.

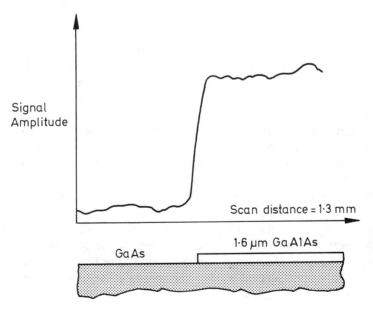

recombination velocity of the order of 10^4 m/s. As a result the injected carrier density remains relatively low. However, when the laser scans onto the GaAs/GaAlAs side of the wafer the signal level increases by about a factor of 10. This can be explained by the following argument: The GaAlAs layer is totally transparent to the incident laser beam and provides no attenuation for black body emission. The injected carriers are therefore created in the GaAs substrate as before and the capping layer presents little modification to the emission characteristics. The carriers are placed close to the materials interface which has an estimated surface recombination velocity of around 500 m/s. As the carrier diffusion lengths are short in GaAs one expects the signal level to vary inversely with the recombination velocity at the dominant surface. The improvement in recombination at the material interface is therefore responsible for the signal increase. This result demonstrates that the technique is able to assess the quality of grown interfaces in a rapid non-contacting manner, and could find use in routine assessment.

4. Effect of Varying Back Surface Emissivity

Images of lightly doped silicon samples taken with PTR often reveal information from their back surface. For instance, scratches and defects on the back surface can show up clearly. The effect can be equally sensitive to material in contact with the back of the wafer. As an example, Fig. 12 shows an image of a silicon wafer taken with the PTR microscope while the sample was resting on a brass nut. The form of the nut is very clearly revealed and appears bright relative to the background. It is expected that the image contrast arises from variations in the surface emissivity of the material at the back surface of the wafer. In order to clarify the mechanism a series of measurements were carried out in which the amplitude of the signal from the wafer was measured for different materials in contact with the back surface. The results of this experiment are shown in Table 1 which shows that the a.c. signal increases with decreasing emissivity of the material on the back surface.

Furthermore, it is noted that if a material with very low emissivity is placed close to, but not in direct contact with, the back surface the output signal is still considerably higher than that corresponding to air backing.

These observations cannot be explained in terms of thermal wave propagation as the sample was too thick to allow penetration by the 10 kHz thermal waves generated in the experiment (the sample thickness was 0.24mm). Also, in these experiments the output signal from the infrared detector was characteristic of those produced by the free carrier emission mechanism - Fig. 3. The image contrast portrayed in Fig. 12 therefore arises predominantly from injected carriers and similarly does not depend upon the carriers reaching the back surface.

Back surface material	PTR signal level	Approx. emissivity
Air	4.1mV	≈ 1
Glass	7.3mV	0.96
Black Rubber	9.6mV	0.86
Brass	20mV	0.05 - 0.22
mirrored surface	42mV	< 0.04

Table 1 Signal level as a function of infrared emissivity for different materials placed at the back surface. Demonstrating possible contrast for Fig. 12.

Fig. 12 PTR image of brass nut taken through a polished silicon wafer.

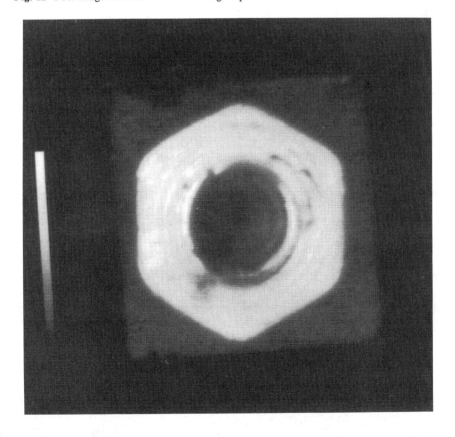

We now consider what happens physically when the modulated laser is incident on a sample of near intrinsic silicon. When the laser is off there are very few free carriers in the wafer so that it is nearly transparent to infrared radiation, with typical lattice absorption coefficients of 1.2 cm^{-1}, thus allowing radiation emitted from the backing material to reach the infrared detector. When the laser is on, the injected carrier density may be as large as 10^{18} cm^{-3}. In this situation free carrier absorption dominates and the wafer becomes comparatively opaque to the infrared radiation emitted from the backing. A greater proportion of the detected infrared radiation is then derived from the silicon sample rather than the backing. The laser has therefore been used to switch the material from transparent to relatively opaque. The signal amplitude detected by the lock-in amplifier is therefore proportional to the difference between the total infrared emission when the laser is on, compared to when it is off.

Following the analysis of section II.3, the infrared radiation leaving the front surface of the semiconductor, when the laser is off, is given by Eq. (13). Assuming that both front and back surfaces of the silicon are polished and possess a reflectivity R_1 and R_2 respectively, the total radiation leaving the front surface can be found from a simple geometrical progression of forward and backward travelling radiation fields, ie.

$$\int_{\lambda_1}^{\lambda_2} \frac{(1-R_1)(1+R_2 T_r)(1-T_r)}{(1-R_1 R_2 T_r^2)} Q_s(\lambda,\theta) \, d\lambda \qquad (25)$$

In addition to the radiation generated within the semiconductor there must be a transfer of radiation between the semiconductor and its environment. We assume that radiation from the surrounding air impinges on the front surface of the semiconductor. A proportion of this radiation is reflected, while the rest enters the semiconductor and undergoes multiple reflections between the front and back surfaces. A fraction of the original radiation finally leaves the front surface of the semiconductor. Furthermore, the back surface of the semiconductor is supported by a backing material of emissivity ε_b. Radiation emitted by the backing material enters the semiconductor and, after multiple reflections, leaves the front surface. To summarize, the total radiation leaving the front surface is therefore the sum of the following components:

$$\int_{\lambda_1}^{\lambda_2} R_1 \, Q_a(\lambda,\theta_a) \, d\lambda \qquad (26)$$

- radiation reflected from front surface;

$$\int_{\lambda_1}^{\lambda_2} \frac{(1-R_1)(1+R_2T_r)(1-T_r)}{(1-R_1R_2T_r^2)} Q_s(\lambda,\theta_s) \, d\lambda \qquad (27)$$

- from within the semiconductor;

$$\int_{\lambda_1}^{\lambda_2} \frac{(1-R_1)^2 R_2 T_r^2}{(1-R_1R_2T_r^2)} Q_a(\lambda,\theta_a) \, d\lambda \qquad (28)$$

- radiation from air leaving front surface, after multiple reflections; and

$$\int_{\lambda_1}^{\lambda_2} \frac{\epsilon_b(1-R_1)(1-R_2)T_r}{(1-R_1R_2T_r^2)} Q_b(\lambda,\theta_b) \, d\lambda \qquad (29)$$

- radiation emitted from backing material, leaving the front surface after multiple reflections.

As in Equation (13) these equations give the emission of infrared radiation from the semiconductor when the laser is on, provided T_r is replaced by T_r^*. The a.c. signal from the infrared detector is the difference in radiation when the laser is switched from off to on, ie.

$$\Delta q = \int_{\lambda_1}^{\lambda_2} \frac{(1-R_1)}{1-R_1R_2T_r^{*2}} [(1-T_r^*)(1+R_2T_r^*) Q_s(\lambda,\theta_s)$$

$$+ (1-R_2) T_r^* \epsilon_b Q_b(\lambda,\theta_b) + (1-R_1) R_2 T_r^{*2} Q_a(\lambda,\theta_a)] \, d\lambda$$

$$- \int_{\lambda_1}^{\lambda_2} \frac{(1-R_1)}{1-R_1R_2T_r^2} [(1-T_r)(1+R_2T_r) Q_s(\lambda,\theta_s)$$

$$+ (1-R_2) T_r \epsilon_b Q_b(\lambda,\theta_b) + (1-R_1) R_2 T_r^2 Q_a(\lambda,\theta_a)] \, d\lambda \qquad (30)$$

Equation (30) is the complete expression defining the a.c. signal level from the infrared detector as a function of the semiconductor optical properties, the temperatures of the surrounding air, semiconductor and backing material, and the emissivity of the backing material. The model assumes that all the components of radiation propagate as plane waves and are normally incident on both the semiconductor surfaces. As a consequence Δq becomes zero when the semiconductor, surrounding air, and backing material are at the same temperature with the backing material emissivity equal to unity, ie. a black body. This is simply a statement of Kirchoff's law, which involves a detailed balance between radiation transfer. In practice, however, the three-dimensional nature of the problem prevents this detailed balance from occurring in such a simple way even when the above conditions apply, as a temperature difference is always established between the point of laser excitation and the rest of the sample/surroundings.

As the sample has been equally polished on both front and back surfaces, the reflection coefficients are assumed to be equal (R, say). Furthermore, the temperature of the air, silicon and backing material are equal. Using the valid approximation that $R^2 \ll 1$ and that the transmission coefficient, T_r, is always close to unity, the output signal, Δq, simplifies to:

$$\Delta q = \int_{\lambda_2}^{\lambda_1} (1 - \varepsilon_b)(1 - R)^2 (T_r - T_r^*) Q(\lambda, \theta) d\lambda \qquad (31)$$

Under these conditions the signal is proportional to $(1 - \varepsilon_b)$ which is broadly supported by the experimental results in Table 1, and accounts for the image contrast in Fig. 12. The effects of bulk properties R, T_r, and T_r^* are enhanced by the factor $(1 - \varepsilon_b)$ which depends only upon the emissivity of the backing material. This mode of operation of the microscope can be likened to conventional infrared imaging of materials where the laser-induced modulations in the transmission coefficient of the silicon sample simulate the action of a mechanical chopper. This imaging mechanism should have applications for subsurface wafer examination and suggests that the signal-to-noise ratio for PTR could be increased by backing the samples with a reflective coating. When performing measurements with the instrument, it is worth noting that some signal contrast could also arise from features at the back surface of the sample and careful interpretation of results may be needed under these circumstances. In general, however, the relative variation of the signal with modulation frequency is not affected.

5. Imaging of Silicon Membranes

This experiment provides an example of conventional thermal wave imaging in heavily doped silicon. The sample consists of a thin silicon membrane provided by Prof. R.M. White, University of California at Berkeley. The sample was prepared by doping a silicon wafer with boron to a concentration greater than 5×10^{19} cm^{-3} within 2μm of the semiconductor surface. A 2mm square window was then exposed on the back surface of the wafer and chemically etched. The wafer had a (100) surface orientation. As

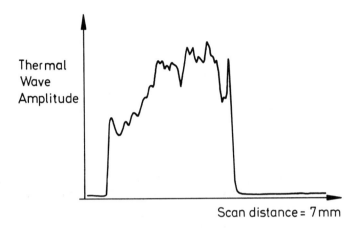
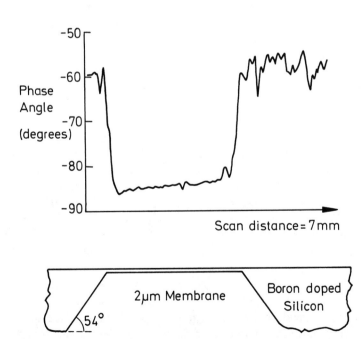

Fig. 13 Silicon membrane sample and corresponding amplitude and phase line scans - showing conventional thermal wave contrast.

the etch rate along the (111) crystal plane is about 30 times faster than along the (100) plane the etching process makes an angle of 54° with the surface. The etch stops when it reaches the boron impurity, thus leaving a 2μm thin membrane. The sample, together with PTR line scans, is shown in Fig. 13.

Over the membrane the high doping density together with the extra surface recombination at the back surface, greatly reduce the injected free carrier emission, while the thermal emission increases dramatically due to the small thermal mass and multiple thermal wave reflections - the thermal diffusion length was set at 55μm. In this case the thermal wave emission dominates. This was confirmed by observing the time response of the infrared emission, which took the form of Fig. 1.

Furthermore, for the thermally thin membrane section simple one dimensional thermal wave analysis gives a surface temperature modulation of:

$$\Delta T = \frac{I_o}{2\rho C \omega d} \quad \text{at -90° phase angle} \tag{32}$$

where ρ is the membrane density, C the specific heat capacity, ω the angular modulation frequency and d the membrane thickness. The dramatic increase in the amplitude over the membrane section is predicted by the thickness dependence of Eq. (32). The expected phase angle change from -45° (thick silicon section) to -90° is also reflected in the line scan result - a more detailed calculation of the predicted phase change would take into account the distributed laser absorption and the three dimensional nature of the PTR instrument.

6. Imaging of Cadmium Mercury Telluride

The ternary alloy cadmium mercury telluride (CMT) is a most valuable photoconductive infrared detector in the 8-14μm wavelength range. The peak spectral response and detectivity depend strongly on the mole fraction of CdTe. For the case of large detector arrays it is therefore essential to ensure compositional uniformity across a series of detecting elements.

At room temperature the band gap of the CMT is small (≈ 0.17eV), giving an intrinsic carrier concentration of around 3×10^{16} cm^{-3} - which is above the Auger recombination threshold for CMT. Normally the semiconductor has to be cooled to 77K in order to sufficiently reduce the carrier concentration to the level required for device operation. Optically injected carriers decay, therefore, via Auger recombination with a considerably shortened lifetime (~ 10ns in some cases) and so the injected carrier densities and free carrier emission are dramatically reduced. The thermal wave emission of infrared radiation dominates, which can be easily verified from the signal dynamics.

Figure 14 shows both optical reflection and PTR images of a single element $Cd_{0.2}Hg_{0.8}Te$ infrared detector. The detector area is 250 x 250 μm^2

with metal contacts shown at the top and bottom of the element. In the thermal wave image the detecting element is brighter than the surrounding metallization and substrate. The modulation frequency was 10 kHz giving a one dimensional thermal diffusion length of ~ 5μm, while the thickness of the CMT element was 11μm. For this case we would expect the signal level to be inversely proportional to the material thermal conductivity. However, the conductivity of $Cd_xHg_{1-x}Te$ is inversely proportional to the mole fraction of CdTe for $x \leq 0.3$. In our case the signal level is thus approximately proportional to the mole fraction of CdTe and some of the image contrast can be attributed to changes in alloy composition. This data could then be related in turn to variations in both electrical and spectral performance of the CMT detector. As the technique is non-contacting and demonstrates good spatial resolution it may find applications in routine measurement of composition during material growth.

V. SUMMARY

This chapter has discussed some of the applications of PTR to semiconductor non-destructive characterization. The technique can be used for both imaging applications and quantitative measurements. We have discussed the trade-off between spatial resolution and the accuracy with which quantitative measurements may be obtained. This has led to the development of two instruments with distinct applications.

For qualitative microscopy, the technique gives results which are similar to those obtained with other techniques such as photoreflectance and photodisplacement, revealing regions of varying damage and contamination. The lateral resolution of the technique can be comparable to these methods but will, in general, be worse since the usable modulation frequencies are lower for reasons of satisfactory signal-to-noise ratio and the response time of the detector (See section III.2). This means that the carriers are not confined to the area under the source, leading to an extended region of enhanced infrared emission. Nevertheless, useful images can be obtained which are rather easier to interpret than those obtained with other techniques. Images showing contamination, lattice and surface damage can be readily produced. Another advantage of the method for on line process monitoring is the fact that the system is robust and simple to align since there is no requirement to focus the pump and probe beams on coincident points.

In order to obtain quantitative accuracy it is necessary to ensure that the area of detection and generation is considerably larger than the diffusion length of the carriers. This naturally means that extremely accurate measurements of carrier properties are obtained at the cost of high lateral resolution. If these conditions are satisfied, the PTR system can be used to perform quantitative measurements of lifetime and recombination velocity.

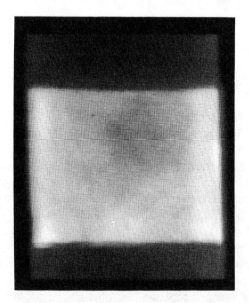

Fig. 14 Optical reflection (top) and thermal wave amplitude (bottom) images of a single $Cd_{0.2}Hg_{0.8}Te$ detector element, showing compositional variations. Image width = 300μm.

The great virtue of the technique compared to other photothermal methods is that the signal can generally be attributed to one mechanism alone. For high quality silicon this arises from the excess blackbody radiation produced by the photogeneration of carriers. This allows accurate quantitative and unique determination of the semiconductor parameters. In other techniques, such as photodisplacement and photoreflectance, the strength of carrier and thermally related effects are comparable which makes quantitative absolute determination of parameters more difficult.

The radiometric instrument thus offers unique advantages for non-destructive semiconductor applications, giving results particularly amenable to quantitative interpretation. It is thus a technique that warrants serious consideration for on-line process control.

VI. REFERENCES

1. R.E. Imhof, *this volume*, Chap. 7.

2. S.J. Sheard and M.G. Somekh, Appl. Phys. Lett. **53**, 2715 (1988).

3. A.C. Tam and B. Sullivan, Appl. Phys. Lett. **43**, 333 (1983).

4. Y. Martin and E.A. Ash, Phil. Trans. Roy. Soc. Lond., **A320**, 257 (1986).

5. W.B. Jackson, N.M. Amer, A.C. Boccara and D. Fournier, Appl. Opt. **20**, 1333 (1981).

6. M.A. Olmstead, N.M. Amer and S. Kohn, Appl. Phys., **A32**, 141 (1983).

7. A. Rosencwaig, in *Photoacoustic and Thermal Wave Phenomena in Semiconductors*, (A. Mandelis, Ed.) 87 (North-Holland, 1987).

8. J.C. White and J.G. Smith, J. Phys., E **10**, 817 (1977).

9. S.J. Sheard, *Photothermal Radiometric Microscopy*, Ph.D. Thesis, (University of London, 1987).

10. G.M. Crean, I. Little and P.A.F. Herbert, Appl. Phys. Lett. **58**, 511 (1991).

11. N. Mikoshiba and K. Tsubouchi, in *Photoacoustic and Thermal Wave Phenomena in Semiconductors*, (A. Mandelis, Ed.) 53 (North-Holland, 1987).

12. K. Cho and C.C. Davis, IEEE J. Quant. Elect. **25**, 1112 (1989).

13. E.A. Ulmer and D.R. Frankl, *Proc. IXth Int. Conf. on Physics of Semiconductors*, 170 (Nauka, 1968).

14. T.S. Moss, Butterworth Scientific Publications, 99 (1959).

15. S.J. Sheard and M.G. Somekh, Elect. Lett., **23**, 1134 (1987).

16. R.N. Bracewell, *The Fourier Transform and its Applications*, (2nd ed.) 110 (McGraw-Hill, 1986).

17. T.M. Hiller, M.G. Somekh, S.J. Sheard and D.R. Newcombe, Mat. Sci. Eng. B, **2**, 107 (1990).

18. G.M. Crean, M.G. Somekh, S.J. Sheard and C.W. See, Mat. Sci. Eng. B, **2**, 207 (1989).

PHOTOACOUSTIC AND PHOTOTHERMAL CHARACTERIZATIONS OF SEMICONDUCTOR SUPERLATTICES AND HETEROJUNCTIONS

Shu-Yi Zhang[*] and Tsuguo Sawada[+]

[*] *Laboratory of Photoacoustic Science, Institute of Acoustics, Nanjing University, Nanjing, The People's Republic of China*

[+] *Department of Industrial Chemistry, Faculty of Engineering, The University of Tokyo, Tokyo, Japan*

I.	INTRODUCTION	152
II.	FUNDAMENTAL FEATURES OF SEMICONDUCTOR HETEROSTRUCTURES	153
	1. Geometric Structures	153
	2. Energy Level Features	155
III.	PHOTOACOUSTIC AND PHOTOTHERMAL SPECTROSCOPIC CHARACTERIZATION OF HETEROSTRUCTURES	157
	1. PA and PT Measurements of Energy Band Structures of III-V Semiconductor Heterostructures	157
	2. Determination of Defects in Amorphous Silicon-Based Heterojunctions by Subgap Absorption	160
	3. Quantum Size Effect in Superlattices	163
	4. Anomalous PA and PT Spectra of Amorphous Semiconductor Superlattices	164
IV.	PHOTO-MODULATED REFLECTANCE CHARACTERIZATION OF HETEROSTRUCTURES	164
	1. Subband States of Quantum Wells	164
	2. Topographical Variations in Barrier Height and Well Width	171
	3. Detection of Two-Dimensional Electron Gas	171
	4. Photoreflectance Mechanisms in MQW	172
V.	THERMAL PROPERTIES OF HETEROSTRUCTURES	172
	1. Thermal Properties of GaAs/AlAs Superlattices	172
	2. Thermal Properties of Amorphous Semiconductor Superlattices	175
VI.	NON-DESTRUCTIVE TESTING OF HETEROSTRUCTURES BY PICOSECOND-LASER-ULTRASONICS	177
VII.	CONCLUSIONS	180
VIII.	REFERENCES	181

I. INTRODUCTION

In recent years, artificial semiconductor superlattices and heterojunctions have motivated intense fundamental and technological interest because of their unusual electronic transport and optical properties [1-6]. Semiconductor superlattices (multi-heterojunctions) are composed of a periodic (or quasi-periodic) sequence of ultra-thin sublayers of alternating composition, of alternating doping or of a combination of alternating composition plus alternating doping. In general, if characteristic dimensions, such as superlattice periods and/or widths of potential wells, are reduced to less than the electron mean free path, the entire electron system will enter a quantum regime of reduced dimensionality with the presence of nearly-ideal hetero-interfaces. The interesting electronic and optical properties are essentially associated with the quantum size effect.

Originally, the compositional crystalline superlattices require a close lattice match of the constituent isotype materials that can be grown epitaxially on top of one another. Otherwise the density of misfit defects usually created at the interfaces is so great that the interesting transport and optical properties will disappear.

However, very recently, Osbourn [7] proposed that the strain associated with the heteroepitaxy of dissimilar materials may itself offer new functionality, since lattice mismatched superlattices can be grown with essentially no misfit defect generation, if the sublayers are sufficiently thin. In this case, the mismatch is accommodated elastically by uniform lattice strain. Therefore, this strain can be a tool for modifying the band structure of semiconductors in a useful and predictable fashion. Since then, a considerable body of work has been devoted to develop the strained heterolayer superlattices due to the more flexible tailorability of their electronic properties, which arises from the competition of quantum-size effects and strain-induced effects.

On the other hand, amorphous semiconductor superlattices were fabricated successfully [8], which extend the range of compositional materials to hydrogenated amorphous semiconductors, that are neither lattice matched nor epitaxial and yet have interfaces which are essentially defect free and nearly atomically sharp. The stringent requirements for lattice matching in crystalline superlattices are relaxed.

The first synthetic semiconductor superlattice was grown by molecular beam epitaxy (MBE) [9], a newly developed thin-film growth technique. To date, a number of different techniques, such as liquid phase epitaxy (LPE) [10], metal-organic vapor phase epitaxy (MOVPE) [11], hot-wall epitaxy (HWE) [12], and plasma chemical vapor deposition (PCVD) [8] etc., have been developed and used to fabricate quantized structures including crystalline and amorphous superlattices, heterojunctions etc.

The structures of semiconductor superlattices and heterojunctions have been studied extensively, with considerable emphasis on the quantum confined states and the associated properties. A wide range of optical methods have been used for these studies, which provide valuable information on the band structures of the materials under consideration. However, their implementation is often involved and, in many cases, difficult. For example, photoluminescence excitation (PLE) spectroscopy [13] and resonant Raman scattering (RRS) measurements [14] are almost always performed at low temperature (typically 4 °K). On the other hand, optical absorption experiments can be performed at room temperature, but a window must be etched in the substrate to allow a sufficient amount of light to be detected. In characterizing microstructural systems, it would be desirable to obtain as much information as possible from a simple measurement preferably at room-temperature.

The photoacoustic (PA) and photothermal (PT) techniques have proven to be sensitive room-temperature probes and have been used for investigating the optical, thermal, mechanical and geometrical properties in macro- or micro-scales of materials [15, 16]. The PA and PT characterizations of semiconductor superlattices, heterojunctions or quantum wells have also attracted wide attention and obtained many interesting and valuable results.

In this chapter, an outline on the progress in PA and PT investigations of semiconductor superlattice and heterojunction structures is presented briefly in order to draw more extensive attention to the diagnostic potential of these techniques.

II. FUNDAMENTAL FEATURES OF SEMICONDUCTORS HETEROSTRUCTURES

1. Geometric Structures

To date, various ultrathin, well-controlled semiconductor heterostructures or superlattices with high quality surfaces and abrupt interfaces have been fabricated widely. As an example, the transmission electron micrograph of an amorphous a-Si:H/a-SiN$_x$:H superlattice deposited by PCVD is shown in Figure 1(a) [17]. From Figure 1(a), the periodic structure can be seen clearly. The interfaces between the sublayers are very clear, indicating that the composition distributions on the two sides of the heterostructures are steep.

On the other hand, the X-ray small-angle diffraction spectrum of the same sample is shown in Fig.1(b), [18]. The diffraction peak in Fig.1(b) shows that the periodicity of the sublayers in the growth process was accurately controlled.

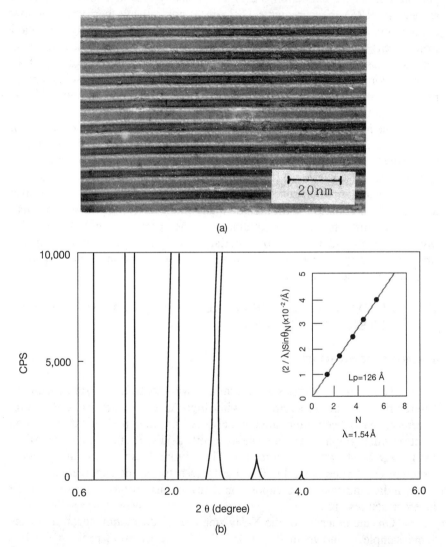

Fig. 1 Geometric structure of an a-Si:H/a-SiN$_x$:H superlattice: (a) TEM micrograph; (From Ref. [17], with permission). (b) Low-angle X-ray diffraction pattern. Insert: Verification of diffraction angles satisfying the Bragg condition. L_p is the period of the superlattice (From Ref. [18], with permission).

2. Energy Level Features

The central feature of superlattices or heterostructures is a one-dimensional periodic potential, where the period d is longer than that of the lattice but short enough that quantum effects are exhibited. For the case of III-V semiconductors, $Al_xGa_{1-x}As/GaAs$ heterostructures have closely matched lattice constants and form a simple super-potential: the GaAs valence band edge is at higher energy and its conduction band edge is at lower energy than the corresponding edges in $Al_xGa_{1-x}As$. Thus the $Al_xGa_{1-x}As$ represents a potential barrier for the electrons in the vicinity of both band edges and then GaAs represents a potential well, as shown in Fig.2(a). The heterostructures with this band-edge configuration are called Type I heterostructures. The heterostructures consisting of InAs and GaSb and their

Fig. 2 One-dimensional periodic potentical models of heterostructures.

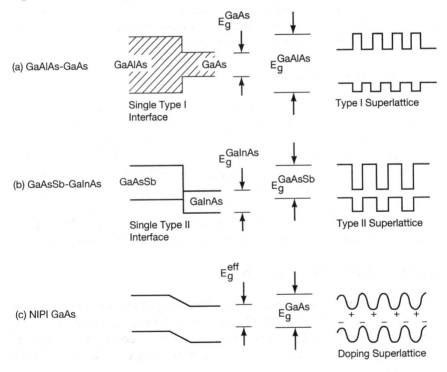

alloys with GaAs have also been synthesized. Because now the conduction band and valence band edges of one semiconductor lie above the corresponding edges of the other, as shown in Fig.2(b), these kinds of heterostructures were called Type II. Furthermore, for alternating doping homojunctions, the energy band structure can be shown as Fig.2(c). Based on the energy band structures, the fundamental properties are summarized as follows:

2.1. One-dimensional Tunnelling Transport Phenomena

The one-dimensional perpendicular transport phenomena across barriers open the way to many fascinating effects. For example, one-dimensional tunnelling phenomena or quantum well wires have been used in tunnelling devices [19]. In addition, oscillations at 200 GHz and nonlinear harmonic generation about 1 THz induced by the resonance tunnelling effect [20] can be implemented, and even the so-called Bloch oscillator [21] should become possible, since the lattice constant of a superlattice is about 10 to 50 times larger than that of a atomic lattice constant.

2.2. Two-dimensional Quantized Structures

In many of these geometries, the quasi-two-dimensional (2D) quantized structures are the basis of optical, electronic and other properties. The free motion of the carriers occurs in only two directions perpendicular to the growth direction, the third direction Z being restricted to a well defined portion of space by momentum, energy, and wave-function quantization.

The most general and surprising feature of the optical properties of quantum wells is the strengthening of the intrinsic optical effects as compared to bulk optical properties. In particular, the quantum efficiency of luminescence has been observed to be much larger in QW structures [5].

2.3. Quantization of Energy Levels

One of the most elementary problems in quantum mechanics is that of a particle confined to a one-dimensional rectangular potential well. If a particle with mass m and wave vector k is completely confined to a layer of thickness L_z by an infinite potential well, then the energies of the bound states are [22,23]

$$E = E_n + (\hbar^2/2m)(k_x^2 + k_y^2) \tag{1}$$

where

$$E_n = (\hbar^2\pi^2/2m)(n/L_z)^2, n = 1, 2, 3 \qquad (2)$$

In reality the potential well is finite and the above solutions are inadequate for a quantitative analysis of the data, but the eigenvalues for a well depth V_0 can be obtained by computer simulation. The quantization of energy levels associated with the confinement of carriers in very thin heterostructures, the so-called quantum size effects, in the case of crystalline superlattices raises the lowest allowed electron and hole energy levels and gives rise to a density of states that increases in discrete steps. This structure in the density of states is reflected in an increase in the optical gap and in discretely spaced absorption peaks.

In amorphous superlattices the fine structure is absent, presumably because of disorder broadening. The evidence for quantum size effects comes from the observation of the large changes in the optical band gap and electrical resistivity.

III. PHOTOACOUSTIC AND PHOTOTHERMAL SPECTROSCOPIC CHARACTERIZATION OF HETEROSTRUCTURES

1. PA and PT Measurements of the Energy Band Structures of III-V Semiconductor Heterostructures

1.1 PA Measurement of $Ga_{1-x}Al_xAs$ on GaAs

Kubota et al.[24] have applied the piezoelectric PZT transducer technique to detect the PA spectra of $Ga_{1-x}Al_xAs$/GaAs heterostructures. A series of $Ga_{1-x}Al_xAs$ layers with different x and about 1.5 μm thickness were grown on GaAs substrates with 300 μm thickness by MBE. The schematic cross section of the sample of a $Ga_{1-x}Al_xAs$ epitaxial layer on a GaAs substrate and the gap energies of the constituent semiconductors are shown in Fig.(3a).

The incident light was illuminated on the epitaxial layer and the PZT transducer was attached to the GaAs substrate. The PA spectra of the samples and substrates are shown in Fig.3(b). From Fig.3(b), it can be seen that the PA spectrum of GaAs only gives the direct band-to-band transition at 1.42 eV, the point *a*. On the other hand, for the samples with epitaxial layers, a decrease around 1.90 eV, point *b*, and the nearly zero signal amplitude around 1.99 eV, point *c*, are observed. This characteristic exciting-photon energy is nearly equal to the indirect transition edge of $Ga_{1-x}Al_xAs$. The broken lines were obtained by reducing the excitation light intensity by the ratio indicated in the figure. This behavior may imply that slightly decreasing the excitation light intensity affects little the PA amplitude due

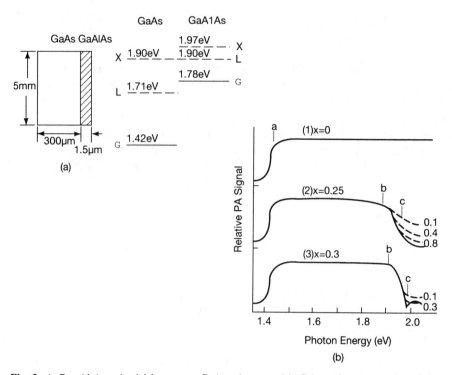

Fig. 3 A $Ga_{1-x}Al_xAs$ epitaxial layer on a GaAs substrate: (a) Schematic cross section of the sample; (b) PAS spectra (From Ref. [24], with permission).

Fig. 4 A $GaAs_xP_{1-x}$ p-n junction on a GaP substrate: (a). Schematic cross section of the sample; (b). PAS spectra (From Ref. [24] with permission).

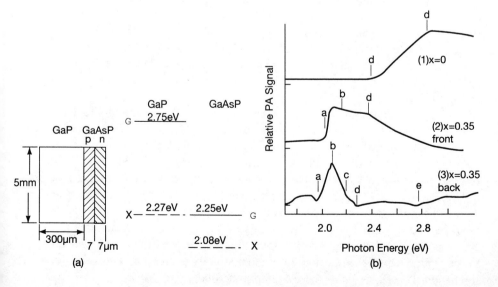

to the direct transition, but obviously changes the amplitude of the indirect-transition part.

In addition, Kanemitsu et al. [25] have used a transparent piezoelectric $LiNbO_3$ transducer to detect the PA amplitude and phase spectra of a $Ga_{1-x}Al_xAs/GaAs$ (x=0.4) heterostructure grown by liquid phase epitaxy (LPE). The PA amplitude spectrum is similar to that of Kubota et al. [26], i.e., the amplitude of the PA signal exhibits a substantial decrease at the indirect edge region and the PA phase is simultaneously changed abruptly. These results illustrate that the process of the PA signal generation from the GaAs substrate differs from that due to the $Ga_{1-x}Al_xAs$ films.

Furthermore, using a spatially resolved electrode array, Kubota et al. [26] found that the PA signal distribution pattern may display the interfacial quality of the heterostructures.

1.2 PA measurement of $GaAs_xP_{1-x}$ p-n Junction on GaP

Kubota et al. [24] have measured the PA spectra of epitaxial layers of p- and n-$GaAs_xP_{1-x}$(x=0.35) with 7 μm thickness each, which were grown on GaP substrates by LPE. The schematic cross section of the sample of a $GaAs_xP_{1-x}$ p-n junction on a GaP substrate and gap energies of the constituent semiconductors are shown in Fig.4(a). In order to contrast, the PA spectra of GaP and of $GaAs_xP_{1-x}$ (x=0.35) of p-n junctions on GaP obtained under different experimental conditions are shown in Fig.4(b). For the GaP substrate, and for photon energies beyond 2.35 eV, the rise of the PA signal is due to the indirect transition absorption, while the direct edge is shown at 2.80 eV. For the epitaxial sample, the obtained spectra are different, as the excitation light is incident on the epitaxial layer or on the GaP substrate, and the transducer is attached to opposite surfaces. The points b and c are the indirect transition and direct transition edges of $GaAs_xP_{1-x}$, respectively. When the incident light is on the GaP side, as shown in the bottom spectrum of Fig.4(b), The PAS signal of GaAsP is obtained on the lower-energy side of GaP, since this region is a forbidden band of GaP. The sharp decreasing point a corresponds to the photon energy of light emission of the junction device.

1.3 Excitonic structures of GaAs and InGaAs single quantum wells (SQW)

Penna et al.[27] have studied the excitonic structures of SQW. Both samples of GaAs and InGaAs SQW were grown by MBE. The GaAs sample consisted of a 104 Å thick layer of GaAs sandwiched between 40 period superlattices of GaAlAs/GaAs, while the InGaAs sample consisted of a single InGaAs layer with 75 Å thickness sandwiched between InAlAs. Because of the quantum confinement effect, the absorption edge of the superlattices occurs at a higher energy and does not interfere with that of the single layer.

The photothermal deflection (Mirage effect) signals obtained from the GaAs and InGaAs samples at room temperature are plotted in Figs.5(a) and 5(b) respectively. Both the absorption spectra show two peaks related to the heavy-hole and light-hole excitons, resulting from the lifting of the degeneracy of the valence band in a two-dimensional system. In the spectrum of InGaAs, the splitting between the peaks is larger than in the case of GaAs because the thickness of the SQW is smaller in the case of InGaAs and also because of differences in the effective masses and the band-gap discontinuities in the two systems. The width of excitonic resonance of the latter is larger than that of the former because alloy fluctuations in the ternary semiconductor are expected to lead to some additional broadening and also due to larger inhomogeneous broadening in a thinner quantum well, induced by larger relative thickness variations. In Fig.5(b) the luminescence spectrum of InGaAs is also shown. Both these observations attest to the high quality of the quantum well.

2. Determination of Defects of Amorphous Silicon-Based Heterojunctions By Subgap Absorption

Traditionally, the optical absorption edge of amorphous semiconductors can be divided into three regions: (a) a power law region; (b) an exponential absorption edge; often called the Urbach edge; and (c) a tail, which extends beyond the Urbach edge into the pseudogap.

The optical absorption of defects and impurities is most readily observed below the band edge, since it is not obscured by the much larger band-to-band absorption. Consequently, subgap absorption spectra should provide information about the number and energy level of defects in these materials. Photothermal deflection spectroscopy (PDS) and photoacoustic spectroscopy (PAS) have been successfully utilized to study the gap states in amorphous thin films.

The number of defects can be estimated by separating the subgap absorption from the exponential band-tail absorption [28]. The excess optical absorption α_{ex} due to gap-state defects is given by

$$\alpha_{ex} = \alpha - \alpha_0 \exp[\hbar\omega/E_0] \tag{3}$$

where α_0 and E_0 are obtained from a fit to the exponential absorption. Then, the number of defects N_s can be calculated from an optical sum rule by the absorption profile [29]

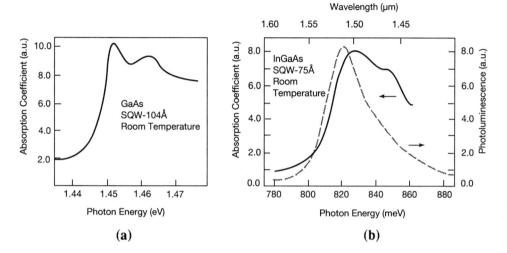

Fig. 5 Absorption spectra of several samples: (a). PDS spectra of a SQW of GaAs; (b). PDS and photoluminescence spectra of a SQW of InGaAs (From Ref. [27], with permission).

$$N_s = \frac{cn_s m_e}{2\pi^2 \hbar^2 e_s^{*2}} \int \alpha_{ex}(E) \, dE \qquad (4)$$

where c is the speed of light, n_s the index of refraction of the amorphous semiconductor, m_e the electron mass and e_s^* the effective charge of the defect. The integration limits extend from zero to the energy at which the exponential absorption terminates.

Asano et al. [30] have measured the defect density of hydrogenated amorphous silicon-based heterojunctions by using PDS. The sample structures of a-$Si_{1-x}C_x$:H/a-Si:H (or a-$Si_{1-x}N_x$:H/a-Si:H) heterojunction with N periods (N=1-10), as shown in Fig.6(a), were prepared by PCVD. The defect density deduced by PDS can be divided into five kinds of defects as denoted by A-E [Fig.6(a)]. The subgap absorption spectra of both samples, in which one is a single-period film of a-$Si_{1-x}C_x$:H(0.2 μm)/ a-Si:H(1.8 μm)(x=0.34) and the other is a film with ten periods of the sublayer a-$Si_{1-x}C_x$:H(0.02 μm)/a-Si:H(0.18 μm), are shown in Fig.6(b). The absorption spectrum for the single period film is well explained by numerically composing the absorption spectra for 1.8 μm thick a-Si:H and 0.2 μm thick a-$Si_{1-x}C_x$:H films. Compared with the absorption spectrum of an undoped a-Si:H film

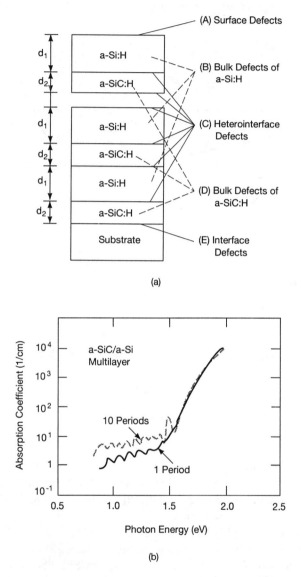

Fig. 6 An a-Si$_{1-x}$C$_x$:H/a-Si:H heterstructure: (a). Sample structure; (b). Subgap absorption spectra (From Ref. [30], with permission).

(2 μm) the Urbach tail is almost the same and the shoulder absorption somewhat increases. The absorption shoulder for the ten-period film is higher than that for the single period film. It has been verified that the increase is not due to the multiple reflection at the a-SiC:H/a-Si:H interfaces, which causes strong and irregular interference fringes at around 1.5 eV. Thus the increase in absorption shoulder is entirely due to the defect absorption at the

a-SiC:H/a-Si:H interfaces.

Asano et al.[30] also investigated the defect densities in a-$Si_{1-x}C_x$:H/ a-Si:H multilayer films having 10 periods and those of samples with a single period as a function of film thickness. The defect density in the single period film slightly increases with the film thickness, which is explained by the increase of the bulk defect in the heterojunction. In contrast, the defect density in a 10 period film shows that the interface-related defects are not located strictly at the interface, but are bulk defects caused by the interfaces, and extended over a range from the interface into the bulk. Meanwhile the defect density in the ten-period a-SiC:H/a-Si:H film increases with increasing a-SiC:H layer thickness. Therefore, the defect density at the a-SiC:H/a-Si:H heterojunction strongly depends on the thickness of a-SiC:H sublayer. Nevertheless, the defect density at the a-$Si_{1-x}N_x$:H/a-Si:H(x=0.50) heterojunction is nearly independent from the a-$Si_{1-x}N_x$:H sublayer thickness.

It must be pointed out that the defect density estimated by Eq.(4) is only for hydrogenated amorphous silicon-based samples, in which the defects are dominatly induced by dangling silicon bonds. For the case of unhydrogenated amorphous silicon, however, such as sputtered samples, Eq.(4) may not be suitable [31].

On the other hand, several authors [32] have analyzed the amplitude of interference fringes in PDS spectra for a-Si:H and amorphous silicon-based alloy thin films in the range of low absorption coefficients. It was shown that the interference fringes are related to the depth profile of the films, e.g. to defects. Thus the analysis provides valuable information about surface and interfacial defects of thin films.

3. Quantum Size Effect in Superlattices

Based on the optical absorption measurement, it has been found that the fundamental energy gaps of semiconductors can be determined. Sai-Halasz [33] investigated the dependence of the fundamental gaps of Type II superlattices on sublayer thicknesses. The fine quantum state structures have been observed. Therefore it is predicted that the quantum size effects can also be studied more easily by the PA and the PT technique.

3.1 GaAs/AlAs Multi-Heterostructures (Superlattices)

Kubota et al.[24] have described the PA spectra of GaAs/AlAs superlattices. GaAs/AlAs superlattices were grown on GaAs substrates of 300 μm thickness by MBE. The superlattices were of variable thicknesses, but the total thickness of multi-heterolayers was not very different for all the samples. The obtained PA spectra are shown in Fig.7(a). The rising feature around 1.42 eV would be due to the direct-transition edge of the GaAs substrate. The characteristic feature of the spectra is a turning point, which shifts to the

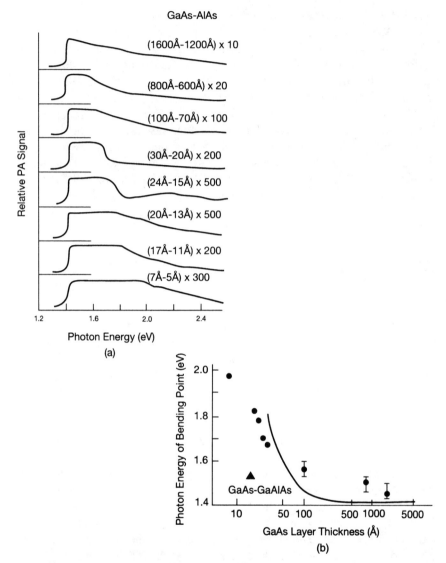

Fig. 7 GaAs-AlAs superlattices: (a). PAS spectra; (b). Dependence of the PAS spectral decrease on GaAs layer thickness (From Ref. [24], with permission).

higher-energy region with decreasing layer thickness as shown in Fig.7(b). Fig.7(b) also shows (solid line) the first quantum level-to-level transition in the GaAs-AlAs multiple quantum well (MQW), which was confirmed by photoluminescence measurements. According to the discussion above, the photon energies of the turning points in PA spectra represent the nature of the indirect-transition. At least, for the samples with thinner layers (~10 Å), it is likely that the indirect transition of a GaAlAs alloy is being observed [34].

3.2 Amorphous Semiconductor Superlattices

Wang et al.[35] firstly presented the investigation of amorphous semiconductor superlattices by photoacoustic spectroscopy. A series of a-Si:H/a-SiN$_x$:H superlattice films and two monolayer films of a-Si:H and a-SiN$_x$:H were prepared and deposited on glass substrates by PCVD. The samples employed were of a fixed thickness (80 Å) of a-SiN$_x$:H and of a range of thicknesses varying from 300 Å to 16 Å of a-Si:H. The total thickness of each sample was approx. 1 μm. The PA spectra of the samples obtained at room temperature by a single-beam photoacoustic spectroscopy are shown in Fig.8(a). For comparison, the optical absorption spectra of the same samples have also been measured by a Beckman DU-8B spectrophoto-

Fig. 8 PAS spectra (a) and optical absorption spectra (b) of a-Si:H/a-SiN$_x$:H superlattices: (A). a-SiN$_x$:H; (B). γ=0.17, T=73; (C). γ=0.35, T=63; (D). γ=0.50, T=55; (E). γ=0.55, T=39; (F). γ=0.79, T=18; (G). a-Si:H; γ: Thickness ratio of the a-Si:H sublayer to superlattice period; T: number of superlattice periods (From Ref. [35], with permission).

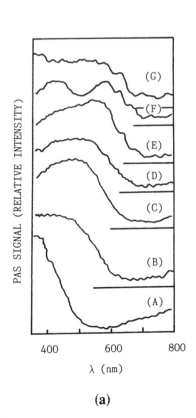
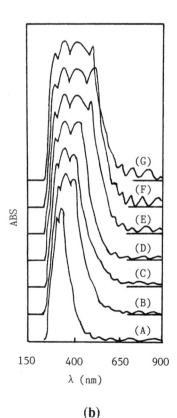

(a)   (b)

meter as shown in Fig. 8(b). The shapes of PA spectra (Fig. 8(a)) are different from the optical absorption spectra (Fig.8(b)), since PAS involves the nonradiative recombination of photogenerated carriers. From this comparison, it can be illustrated that most of the photogenerated carriers recombine radiatively. From Figure 8, it is very clear that the absorption edge of the PA spectrum shifts to a shorter optical wavelength with decreasing a-Si:H thickness, which corresponds to the potential well width of the multilayer film. This is the so called "blue shift" phenomenon of the absorption edge associated with the quantum size effect.

Using Tauc's equation

$$(\alpha \hbar \nu)^{1/2} = B(\hbar \nu - E_g) \qquad (5)$$

where α is the absorption coefficient, $h\nu$ is the incident photon energy, B is a constant and E_g is the optical energy gap. From Fig.8(b) it is easy to obtain the curves of $(\alpha \hbar \nu)$ vs $\hbar \nu$ as shown in Fig.9. In Fig.9, the intersections of the dotted lines with the horizontal axis represent the optical gaps E_g of the corresponding samples. This can be interpreted that the gap E_g of a-Si:H/a-SiN$_x$:H superlattices should increase by different values with decreasing widths of the wells.

Fig. 9 Dependence of $(\alpha h\nu)^{1/2}$ on incident photon energy of a-Si:H/a-SiN$_x$:H superlattices (From Ref. [35], with permission).

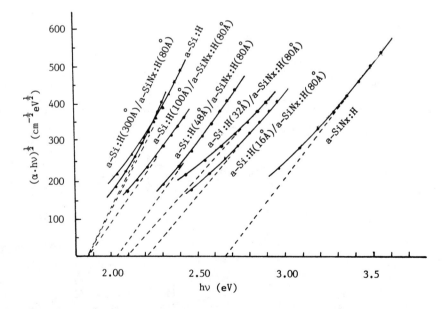

4. Anomalous PA and PT Spectra of Amorphous Semiconductor Superlattices.

In Fig.8(a), one can see that an anomalous "valley" exists in the spectrum labelled F. In order to further study the anomalous characteristics of amorphous semiconductor superlattices, another set of a-Si:H/a-SiN$_x$:H (x~1) multilayer films with the thickness of each film a few micrometers were deposited on glass substrates by PCVD, which has twenty periods (T=20) but different thickness ratio γ of the a-Si:H sublayer relative to the period of the superlattices. Both PA and PT spectroscopies have been used to study the superlattices and have obtained the same results, which are shown in Fig.10(a) [36]. In addition, optical absorption spectra were obtained as shown in Fig.10(b).

From Figure 10, it can be observed that "blue shift" phenomena induced by the quantum size effects disappear, since the samples have wider potential wells. But the anomalous absorption "valley" appears in each PA spectrum at the optical wavelength range of 500-600 nm. However, in the optical absorption spectra, Fig.10(b), no "valley" can be found. Apparently, the anomalous "valley" cannot be explained by the optical interference of multiple reflections from a-Si:H/a-SiN$_x$:H interfaces. Therefore, an anomalous radiative recombination of photogenerated carriers probably occurs in the region where the "valley" appears in the PA (or PT) spectra.

Furthermore, it is interesting to point out that, for these special samples, anomalous elasticity, the so-called "phonon softening" [37], and anomalous thermal diffusivity have been found [38].

IV. PHOTO-MODULATED REFLECTANCE CHARACTERIZATION OF HETEROSTRUCTURES

The Photoreflectance (PR) technique [39] is a contactless mode of electromodulation reflectance which has been used successfully to characterize semiconductors. The PR technique has recently been applied to characterize alloy composition, quantum well width, and interfacial quality of multiple quantum wells (MQW) etc. [40]. In PR the modulation of the optical constants is produced by photo-injection of electron-hole pairs by a modulated pump light source. Some of the main investigated results are now described:

1. Subband States of Quantum Well Structures

The nature of subband states in both the conduction and valence bands of MQW structures and single heterojunctions has been studied with intense theoretical and experimental interest. In the study of Glembocki et al.[40], the samples of MQW structures consisted of ~ 30 periods of GaAs/Al$_x$Ga$_{1-x}$As layers grown on semi-insulating GaAs substrates and clad with either 200 Å-

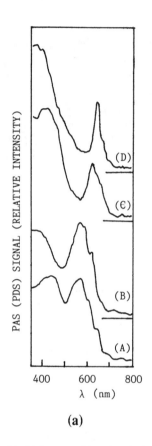 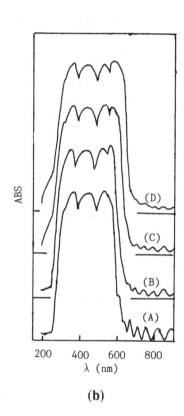

Fig. 10 PAS spectra (a) and optical absorption spectra (b) of a-Si:H/a-SiN$_x$:H superlattices with 20 T: (A) γ=0.92; (B) γ=0.8; (C) γ=0.7; (D) γ=0.6 (From Ref. [36], with permission).

thick GaAs or 1000 Å thick Al$_x$Ga$_{1-x}$As. The Al$_x$Ga$_{1-x}$As layers separating the quantum wells were about 125 Å thick. The photomodulation was accomplished by mechanically chopping visible light from a He-Ne laser at 150 Hz. The PR spectra of single GaAs/Al$_{0.7}$Ga$_{0.83}$As heterojunction and GaAs/Al$_x$Ga$_{1-x}$As MQW samples obtained at room temperature are shown in Fig. 11. In Figure 11, the topmost trace is a PR spectrum for single undoped GaAs/Al$_x$Ga$_{1-x}$As heterojunction, covering the spectral range between the direct gaps of the GaAs and Al$_x$Ga$_{1-x}$As components, i.e., the $E_0(\Gamma_{8,v} \rightarrow \Gamma_{6,c})$ transition of both materials. Both lineshapes result from E_0 transitions of GaAs and Ga$_{0.83}$Al$_{0.17}$As. The other three traces of Fig.11 are PR spectra of the MQW samples. It can be seen that each spectrum contains a weak GaAs PR line near 1.42 eV which originated from the GaAs buffer layer. It can also be seen that each spectrum exhibits a line near 1.70 eV corresponding to the

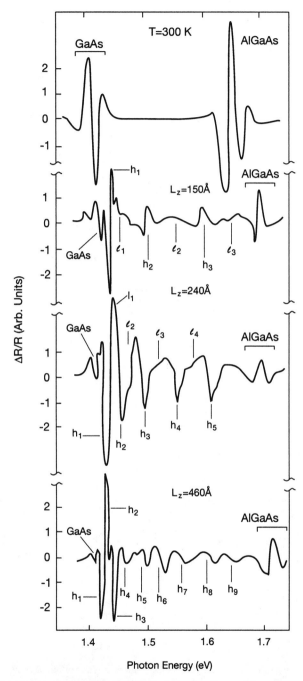

Fig. 11 PR spectra of an undoped GaAs/AlGaAs heterojunction and three MQW samples with x=0.2 (From Ref. [40], with permission).

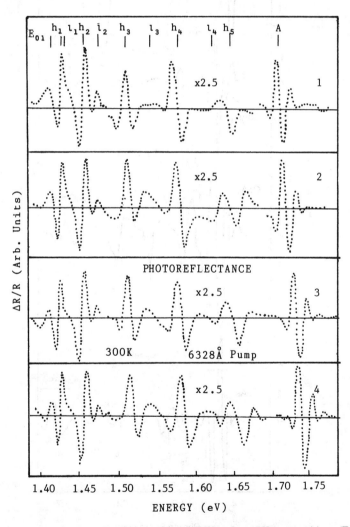

Fig. 12 PR spectra of a GaAs/GaAlAs (x≈0.24) MQW at four different positions (From Ref. [42], with permission).

E_0 transitions in the $Al_xGa_{1-x}As$, allowing one to obtain x for each sample [41].

The strongest features in the spectra are the peak increase in number, accompanied by energy separation decrease, with increasing quantum well width. This is characteristic of quantum well interband transitions. To identify all of the PR lines Glembocki et al.[40] have performed a finite square well calculation of the subband structure. The calculated results are marked h_1, h_2, ... h_n and ℓ_1, ℓ_2, ... ℓ_n, which denote the interband transitions from heavy (h_1, ... h_n) and light (ℓ_1 ..... ℓ_n) hole states to conduction subband states.

2. Topographical Variations in Barrier Height and Well Width

Parayanthal et al. [42] reported the use of room temperature PR to evaluate the topographical variations in barrier height (alloy composition) and quantum well width across the surface of a GaAs/GaAlAs MQW sample with a spatial resolution of approx. 100 μm.

The GaAs/GaAlAs (x=0.24) MQW sample was grown by MBE and mounted on an x-y stage. The pump beam, from a 1 mW He-Ne laser, was defocused to cover an area somewhat larger than the region to be scanned. At this low power level the spectra were independent of power density.

A set of experimental PR spectra obtained at various positions on the sample is shown in Fig.12. The spectra exhibit features from all confined quantum levels as well as a peak related to the band gap of the GaAlAs barrier layers. By employing a least-squares fit of the experimental data to the Aspnes third-derivative lineshape function for electromodulation spectra [43], the quantum-well width and barrier height can be obtained. From Figure 12 one finds that there is a variation not only in A, which is related to the direct gap of the GaAlAs barrier layer, but also in the well width in going from position 1 to 4. Making use of a detailed lineshape fit, variations in quantum width as small as one monolayer can clearly be detected. It is possible to evaluate variations in barrier height to within several meV and in well width to within 2 Å. This method as a non-destructive depth profilometry is extremely convenient compared with other characterization techniques.

3. Detection of Two-Dimensional Electron Gas

Glembocki et al.[40] utilized GaAs/AlGaAs modulation-doped heterojunctions to demonstrate two-dimensional electron confinement effects. A selectively n-doped $Al_xGa_{1-x}As$ layer introduces a large density of electrons (> 10^{11} cm^{-2}) at the GaAs side of the interface. This results in the formation of a triangular quantum well and a corresponding subband ladder. For the samples which exhibited a two-dimensional electronic gas (2DEG), the PR spectra have extra features compared with those of non-2DEG samples. This suggests that the presence of a 2DEG in a modulation-doped heterojunction can be detected from a relatively simple room-temperature measurement, which is generally otherwise detected by cyclotron resonance measurements performed at 4.2 K.

4. Photoreflectance Mechanisms in MQW

Although room-temperature PR measurements have been shown to be valuable for the characterization of the alloy composition, quantum well width and interfacial quality of MQW, the nature of the modulation mechanisms that give rise to the PR effect remains poorly understood. For example, there is an apparent discrepancy between recent studies that a PR spectrum is dominated by non-excitonic interband transitions in bulk materials, and the wealth of investigations indicating that the interband optical spectrum of MQW is dominated by excitonic transitions. Shanabrook *et al.* [44] performed PR measurements in conjunction with PLE, which may allow the mechanisms to be determined. The study indicates that the modulation of the interband excitonic transitions, rather than the band to band transitions, is observed in the PR spectrum.

Furthermore, Zheng *et al.* [45] have obtained some photoreflectance results on a $Ga_{1-x}Al_xAs/$ GaAs quantum well at 4.2 K. According to their reflectance results of the same single quantum wells with different barrier layer thicknesses, they suggested that the interference between the reflected waves from air/$Ga_{1-x}Al_xAs$ surface and from quantum well interfaces also contribute to the observed photoreflectance lineshapes. Comparing the experimental results to calculations, the absorption coefficient for the exciton transition can be determined.

V. THERMAL PROPERTIES OF HETEROSTRUCTURES

As described in sections III and IV, the optical properties of semiconductor heterostructures have been extensively studied. The electrical properties of heterostructures have been studied much more widely. However, there have been just a few reports on the thermal properties of these MQW materials, although thermal properties are very important when one designs device structures.

1. Thermal Properties of GaAs/AlAs Superlattices

Yao [46] first presented the use of an a.c. calorimeter [47] to measure the thermal properties of superlattices. The sample was partially irradiated with chopped light and a mask was driven by a micrometer to produce a change in the a.c. temperature amplitude (T_{ac}). A thermocouple wire was attached to the surface of the sample to measure T_{ac}. The GaAs/AlAs

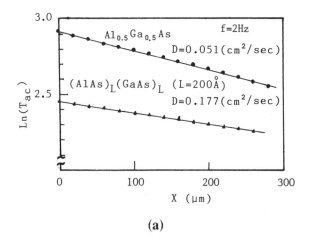

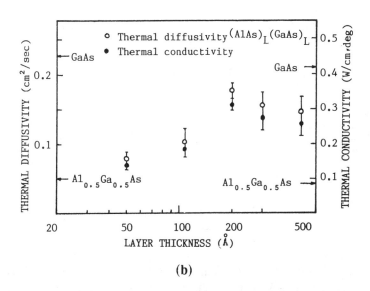

Fig. 13 Thermal properties of AlAs/GaAs superlattices: (a). a.c. temperature vs. x; (b) Thermal diffusivity and conductivity vs. layer thickness (From Ref. [46], with permission).

superlattices were grown on (100) GaAs substrates by MBE. The thicknesses of each GaAs and AlAs layer were the same and ranged from 50 to 500 Å. A typical thickness of the superlattices was 10 μm. The results of the measurements are shown below.

1.1 Thermal diffusivity

The relation of T_{ac} and the distance (x) between the edge of the mask and the thermocouple junction can be approximately represented as

$$\log(T_{ac}) = \log(\sqrt{2}\,Q/K) - Kx \tag{6}$$

where $K = (\pi f/D)$, D being the thermal diffusivity, and K and Q are thermal conductivity and the amplitude of a.c. heat generated by the chopped light, respectively. Obviously, the measured thermal properties by this method are associated with heat flow along the layer. The measurements were performed at room temperature. The GaAs substrate on which the superlattice was grown was etched off by using a solution of NH_4OH and H_2O_2, so that the measurement was carried out without any substrate influence.

Figure 13(a) shows the plots of T_{ac} vs. x for an $Al_{0.5}Ga_{0.5}As$ alloy and the AlAs/GaAs superlattice with L = 200 Å. Evidently, the K value of the $Al_{0.5}Ga_{0.5}As$ alloy sample is larger than that of the superlattice. Figure 13(b) shows the thermal diffusivity values of superlattices plotted against the superlattice sublayer thickness. The measured thermal diffusivities of $Al_{0.5}Ga_{0.5}As$ and GaAs are also shown by arrows. It is noted that the measured thermal diffusivities of the superlattices lie in between those of the GaAs and $Al_{0.5}Ga_{0.5}As$ alloy. The thermal diffusivity decreases as the layer thickness decreases and seems to eventually approach that of the AlGaAs alloy in the limit of the monolayer superlattice.

When AlAs and GaAs layers in AlAs/GaAs superlattice are sufficiently thick, the thermal properties of both AlAs and GaAs layers would contribute independently to the thermal properties of the superlattice. In this case, the superlattice will be regarded as a composite material made of AlAs and GaAs. Then the thermal diffusivity of the superlattice will be given by the weighted average of thermal diffusivities of AlAs and GaAs, that is [46]

$$D = \frac{c_1 \rho_1 d_1 D_1 + c_2 \rho_2 d_2 D_2}{c_1 \rho_1 d_1 + c_2 \rho_2 d_2} \tag{7}$$

where c_1 (or c_2) is the specific heat, ρ_1 (or ρ_2) is the density, d_1 (or d_2) is the layer thickness and 1 and 2 denote GaAs and AlAs, respectively. The calculated thermal diffusivity of this case is much larger than the measured values of AlAs/GaAs superlattice in the described experiments. This fact indicates that a composite material of AlAs and GaAs may be realized with much thicker superlattice spacing than 500 Å.

1.2. Thermal Conductivity

The thermal conductivity can be derived from the thermal diffusivity. The obtained thermal conductivities of the superlattice are also shown in Fig.13(b). It is noted that the thermal conductivity of GaAs/AlAs superlattices is larger than that of the $Al_{0.5}Ga_{0.5}As$ alloy. This is obviously due to the suppression of the alloy scattering in superlattices, in which AlAs and GaAs layers are coherently arranged. Also noted is a decrease in the thermal conductivity with a decrease in the superlattice period. A weighted average of thermal conductivity of GaAs and AlAs corresponds to that of a composite material composed of GaAs and AlAs. This value is almost identical with that of a virtual crystal of $Al_{0.5}Ga_{0.5}As$ having an averaged atomic mass and atomic radius of an $Al_{0.5}Ga_{0.5}$ atom. Since the measured thermal conductivity values of the superlattices are smaller than this value, excess scattering, such as interface scattering, may still exist in the superlattices.

1.3. Thermal Resistivity

The comparison of the thermal resistivity values of the superlattices with the reported values of $Al_{0.5}Ga_{0.5}As$ alloy has also been undertaken [46]. It was found that the thermal resistivities of superlattices are smaller than the alloy [48]. Therefore, it is expected that the thermal properties of semiconductor devices will be improved by using superlattices instead of alloys.

2. Thermal Properties of Amorphous Semiconductor Superlattices

2.1 Thermal diffusivity of amorphous superlattices

Zhang et al.[49] presented a study of the thermal diffusivity of the amorphous semiconductor superlattices a-Si:H/a-SiN$_x$:H using the mirage detection technique [50]. In the measurement, it was assumed that the samples were opaque but the substrates were transparent to the incident optical beam. Therefore, it was necessary to choose an optical heating source whose wavelength is strongly absorbed by the sample but highly transparent through the substrate.

For the 458 nm beam of Ar$^+$ ion laser, all of the superlattice and a-Si:H films can be considered as opaque materials, i.e. the light penetration depth is much less than the thickness, but the a-SiN$_x$:H film is almost transparent in the visible region.

According to the one-dimensional treatment [50], the phase lag between rear and front surface excitations is obtained as follows:

$$\tan(\Delta\phi) = \frac{1 - R_{ab}\exp(-2L_s/\mu_s)}{1 + R_{ab}\exp(-2L_s/\mu_s)} \tan(L_s/\mu_s)$$

and

$$R_{sb} = \frac{e_s - e_b}{e_s + e_b} \qquad (9)$$

where L_s and μ_s are the thickness and the thermal diffusion length, respectively, of the sample. e_s and e_b are the thermal effusivities of the sample and the substrate, respectively.

For a set of a-Si:H / a-SiN$_x$:H superlattices, it was found that the phase lag $\Delta\phi$ changes proportionally to the square root of the frequency over the frequency range of 100 Hz-40 kHz, which indicates $e_s = e_b$, i.e., $R_{sb} = 0$ within the experimental error. Then

$$\frac{\Delta\phi}{f^{1/2}} \propto L_s (\pi/D)^{1/2} \qquad (10)$$

Based on the experimental results, the deduced thermal diffusivities of the samples are shown in Fig.14. Evidently, the measured thermal diffusivity is in the perpendicular direction of the films.

In order to explain the experimental results, a theoretical model consisting of thermal resistors in series [51] was used in Ref [49]. An expression for the effective thermal conductivity K_s of the superlattices may then be written as

$$1/K_s = \gamma/K_1 + (1-\gamma)/K_2 \qquad (11)$$

where K_1 and K_2 are the thermal conductivities of the constituent sublayers a-Si:H and a-SiN$_x$:H respectively. The effective thermal diffusivity is given by

$$D_s = \frac{K_s}{\gamma \rho_1 C_1 + (1-\gamma)\rho_2 C_2} \qquad (12)$$

Fig.14 also shows the theoretical results which are essentially in agreement with the experimental values. From the theoretical curve, the thermal conductivity ratio of both constituents may then be deduced.

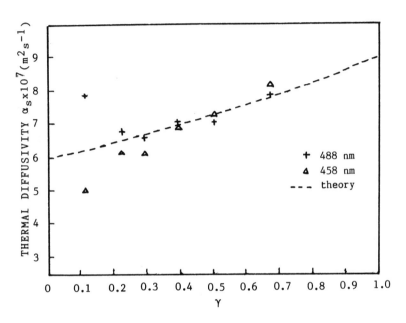

Fig. 14 Thermal diffusivity vs. the thickness fraction γ of a-Si:H related to the period of a-Si:H/a-SiN$_x$:H superlattices (From Ref. [49], with permission).

2.2. Anomalous Thermal Diffusivity of Amorphous Superlattices

For another set of a-Si:H/a-SiN$_x$:H (x~1) superlattices, which displayed anomalous optical and elastic properties, it was found that the experimental values of thermal diffusivity are much larger than the theoretical results, [38]. It was thus demonstrated that the diffusivity of these kinds of amorphous superlattices cannot be calculated by the theory described above, and that the mechanisms of these anomalous properties cannot be explained clearly yet.

VI. NON-DESTRUCTIVE TESTING OF HETEROSTRUCTURES BY PICOSECOND LASER-ULTRASONICS

The first application of picosecond laser pulses to time-resolved measurements was reported by Shelton and Armstrong [52]. Since then a number of techniques have been developed to excite, probe, delay, gate and synchronize picosecond pulses.

The simplest version of the experimental technique is shown schematically in Fig.15(a). A short light pulse is incident on an optically-absorbing film on a substrate. The temperature of the heated part of the film rises, setting up a thermal stress in the film. The increase in temperature at a distance z below the surface is [53]

$$\frac{(1-R)Q_e}{A_0 \xi C_s} e^{-z/\xi} \qquad (13)$$

where R is the optical reflectivity, Q_e is the energy in the light pulse, A_0 is the area of the film which is excited, ξ is the absorption length for the light in the material, and c_s is the specific heat. If the thickness of the film is much greater than the absorption length, the stress launches an elastic pulse into the film and it can be shown that the strain associated with this pulse is [54]

Fig. 15 Picosecond-laser-ultrasonic experiment: (a) Schematic diagram of the experiment; (b). Photoinduced reflectivity changes in an a-Ge/a-Si superlattice (From Ref. [55], with permission).

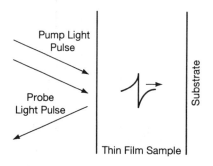

(a)

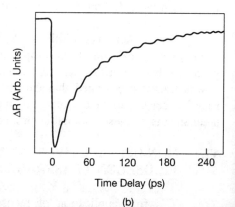

(b)

$$\eta(z,t) = -\frac{(1-R)Q_e\beta}{2A_0\xi C_s}\left(\frac{1+\nu}{1-\nu}\right)e^{-|Z-\upsilon t|/\xi}F(Z-\upsilon t) \tag{14}$$

where β is the thermal expansion coefficient, ν is the Poisson ratio, υ is the acoustic velocity and the function F determines the spatial shape of the pulse. After propagation through the film, the acoustic pulse is partially reflected at the film-substrate interface. When this echo returns to the free surface of the film, the elastic strain produces a change in the optical properties near the surface, and as a result, a small change in the optical reflectivity ΔR occurs. Then a probe beam separated from the same pulsed laser is used to detect the reflectivity change. ΔR can be expressed in the form [54]

$$\Delta R(t) = \int f(z)\eta(z,t)dz \tag{15}$$

where f(z) is called the sensitivity function,

$$f(z) = f_0\cos(\frac{2\pi n_s j}{\lambda}-\varphi)e^{-z/\xi} \tag{16}$$

where f_0 and ϕ are constants dependent upon n_s and ξ.

Grahn et al. [55] have utilized a short light pulse with width 0.1 ps, energy per pulse 0.2 nJ, and repetition rate 108 MHz to study in the time domain the vibrations of a multilayer structure composed of Ge and Si. When the light pulse is absorbed in the multilayer, a stress distribution is set up and the structure is set into vibration. According to the simplest picture of this process, the absorption of light raises the temperature and excites carriers. Therefore, both the thermal expansion and the electronic contribution induce stresses. Because of the different optical properties of Ge and Si, the spatial dependence of the initial stress distribution is rather complicated.

A typical result of the response $\Delta R(t)$ for a multilayer is shown in Fig.15(b). It consists of two superimposed terms. The first term is a rapid decrease of ΔR immediately after t=0 followed by a decay. The second term consists of a weakly damped oscillation of frequency 91.3 GHz together with some irregular oscillatory components which persist only for a short time. The first term is the result of the change in optical properties due to photo-excited carriers. The second term (oscillatory component) is clearly a result of the vibration of the multilayer structure. Grahn et al.,[55] have also analyzed the results in terms of the spectrum of normal modes of the multilayers and have shown that there exist surface modes at the free surface of the multilayer. The detailed theories can be found in the related references.

The sound velocity and attenuation, as well as the index of refraction and microstructure of the films can be determined, by the measurement of the propagation time of the strain pulse and the period of the oscillations, with knowledge of the incident angle.

Harata et al.,[56] by using a two-crossed picosecond laser beam technique, have reported, surface acoustic waves in the GHz frequency range and related photothermal surface gratings in thin films and bulk materials. Besides the studies of acoustic properties and the index of refraction, the thermal diffusivity of materials can thus be estimated by the relaxation time of the photothermal grating, which is formed as a non-propagating component of thermo-elastic waves.

VII. CONCLUSIONS

We have summarized briefly the fruitful research activities of PA and PT non-destructive characterizations of semiconductor superlattices and heterojunctions after a simple introduction of the fundamental features of semiconductor heterostructures. The main results to-date can be concluded as follows:

(1) PA and PT spectroscopies are very useful techniques for characterizing the energy-band structures and subgap absorptions of semiconductor heterostructures because of their high sensitivity.

(2) The PR method is sensitive to the electric fields and, occasionally to the temperature field on the surface. PR spectroscopy has been used to evaluate the alloy mole-fraction x in semiconductors, the quality of thin films and interfaces between layers, as well as topographical variations in barrier height and quantum well width of MQW, etc.

(3) The PT technique is a method developed to study the thermal properties of thin films including multilayers in non-destructive and non-contact configurations, which cannot always be investigated by other techniques.

(4) In particular, owing to the recent development of picosecond optics, picosecond pulsed laser-ultrasonics will develop rapidly to detect ultrathin films and other microstructures with high temporal and spatial resolutions, through the excitation of coherent, high-frequency phonons.

Semiconductor superlattices and heterojunctions are new materials developed rapidly for their important applications in microelectronic science and industry. PA and PT techniques have proven to be useful tools to characterize the optical, electronic, thermal and elastic properties of the heterostructures, which usually cannot be easily obtained by other techniques. However, the mechanisms have not been understood clearly, because of the complex nature of multi-layered structures. This chapter essentially gives the description of some observed phenomena, with no emphasis on theoretical explanation. Further experimental and theoretical studies will be necessary in the future to yield a quantitative understanding of these phenomena.

VIII. REFERENCES

1. A.G. Milnes and D.L. Feucht, *Heterojunctions and Metal-Semiconductor Junctions* (Academic, N.Y., 1972).

2. K. Ploog and G.H. Donler, Advances in Physics, **32**, 285 (1983).

3. B. Abeles and T. Tiedje, in *Semiconductors and Semimetals,* **Vol.21c**, Chap.12, (Academic, New York, 1984).

4. R. Dingle, *Semiconductors and Semimetals*, **Vol. 24**, (Academic, New York, 1987).

5. T.P. Pearsall, *Semiconductors and Semimetals*, **Vol. 32** (Academic, New York, 1990).

6. G. Allan, G. Bastard, N. Boccara, M. Lannoo and M. Voos, *Heterojunctions and Semiconductor Superlattices*, (Springer-Verlag, Berlin, 1985).

7. G.C. Osbourn, (1982), J. Appl. Phys., **53**, 1586 (1982).

8. B. Abeles and T. Tiedje, Phys. Rev. Lett., **51**, 2003 (1983).

9. L.L. Chang, L. Esaki, W.E. Howard and R. Lucleke, J. Vac. Sci, Technol. **10**, 11 (1973).

10. E.A. Rezek, N. Jr. Holonyak, B.A. Vojak, G.E. Stillman, J.A. Rossi, D.L. Keune and J.D. Fairing, Appl. Phys. Lett., **31**, 289 (1977).

11. S. Hersee and J.P. Duchemin, Ann. Rev. Mater. Sci., **12**, 65 (1982).

12. H. Kinoshita and H. Fujiyasu, J. Appl. Phys., **51**, 5845 (1981).

13. R.C. Miller and D.A. Kleinman, J.Lumin., **30**, 520 (1985).

14. G. Abstreiter and K. Ploog, Phys. Rev. Lett., **42**, 1308 (1979).

15. A. Rosencwaig, *Photoacoustics and Photoacoustic Spectroscopy*

(Wiley, New York, 1980).

16. A.J. Sell, *Photothermal Investigations of Solids and Fluids*, (Academic, New York, 1988).

17. K.J. Chen, G.Z. Mao, H. Chen, J.F. Du and X.R. Zhang, Thin Solid Films, **163**, 55 (1988).

18. K. Chen, J. Xu, L. Zhou, J. Jian, Z. Li and J. Du, Chinese Phys. Lett., **8**, 432 (1991).

19. N. Yokoyama, K. Imamura, T. Ohshima, H. Nishi, S. Muto, K. Kondo and S. Hiyamizu, Jpn. J. Appl. Phys., **23**, L311 (1984).

20. L.L. Chang, L. Esaki and R. Tsu, Appl. Phys. Lett., **24**, 593 (1974).

21. R.O. Grondin, W. Porod, J. Ho, D.K. Ferry and G.J. Iafrate, Superlattices Microstruct., **1**, 183 (1985).

22. A. Messiah, *"Quantum Mechanics"*, Vols. I, II (North-Holland, Amsterdam, 1961).

23. R. Dingle, W. Wiegmann and C.H. Henry, Phys. Rev. Lett., **33**, 827 (1974).

24. K. Kubota, H. Murai and H. Nakatsu, J. Appl. Phys., **55**, 1520 (1984).

25. Y. Kanemitsu and H. Nabeta, Appl. Phys. Lett., **59**, 715 (1991).

26. K. Kubota and H. Murai, J. Appl. Phys., **56**, 835 (1984).

27. A.S.F. Penna, J. Shah and A.E.D. Giovanni, Appl. Phys. Lett., **47**, 591 (1985).

28. W.B. Jackson and N.M. Amer, Phys. Rev. **B25**, 5559 (1982).

29. M.H. Brodsky, M. Cardona and J.J. Cuomo, Phys. Rev. **B16**, 3556 (1977).

30. A. Asano, T. Ichimura, Y. Uchida and H. Sakai, J. Appl. Phys., **63**, 2346 (1988).

31. E. Bustarret, D. Jousse, C. Chaussat and F. Boulitrop, J. Non-Cryst. Solids, **77&78**, 295 (1985).

32. G. Amato, G. Benedetto, L. Boarino and R. Spagnolo, Appl. Phys. A **50**, 503 (1990); G. Amato, G. Benedetto, F. Fizzotti, G. Manfredotti and R. Spagnolo, Phys. Stat. Sol. **119**, 169 (1990); A. Asano and M. Stutzmanu, J.Appl.Phys., **70**, 5025 (1991).

33. G.A. Sai-Halasz, in *Physics of Semiconductors*, (B.L.Wilson, Ed.) 21 (Institute of Physics, London, 1981).

34. J.P. Van der Ziel and A.C. Gossard, J. Appl. Phys., **48**, 3018 (1977).

35. Z.C. Wang, S.Y. Qiu and S.Y. Zhang, in *Photoacoustic and Photothermal Phenomena*, (P. Hess and Pelzl, Eds.) 232 (Springer-Verlag, Berlin, 1987).

36. S.Y. Zhang, B.F. Zhang, Z.C. Wang, E.A. Gardner and J.P. Zhou, in *Photoacoustic and Photothermal Phenomena II*, (J.C. Murphy, J.W. Maclachlan Spicer, L.C. Aamodt and B.S.H. Royce, Eds.) 177 (Springer-Verlag, 1990).

37. Q. Shen, S.Y. Zhang, Z.C. Wang, Z.N. Lu and J. Yu, Phys. Rev. **B39**, 11016 (1989).

38. S.Y. Zhang, A.C. Boccara, D. Fournier, J.P. Roger and Z.C. Wang, in *Rev. Prog. Non-destructive Eval.*, **9**, 1129 (1990).

39. J.L. Shay, Phys. Rev., **B2**, 803 (1970).

40. O.J. Glembocki, B.V. Shanabrook, N. Bottka, W.T. Beard and J. Comas, Appl. Phys. Lett., **46**, 970 (1985).

41. O. Berolo and J.C. Woolley, Can. J. Phys., **49**, 1335 (1971).

42. P. Parayanthal, H. Shen, F.H. Pollak, O.J. Glembocki, B.V. Shanabrook and W.T. Beard, Appl. Phys. Lett., **48**, 1261 (1986).

43. H. Shen, P. Parayanthal, F.H. Pollak, M. Tomkiewiez, T. Drummond and J.N. Schulman, Appl. Phys. Lett., **48**, 653 (1986).

44. B.V. Shanabrook, O.J. Glembocki and W.T. Beard, Phys. Rev. **B35**, 2540 (1987).

45. X.L. Zheng, D. Heiman and B. Lax, Appl. Phys. Lett., **52**, 287 (1988).

46. T. Yao, Appl. Phys. Lett., **51**, 1798 (1987).

47. I. Hatta, Y. Sasuga, R. Kato and A. Maesono, Rev. Sci. Instrum., **56**, 1643 (1985).

48. M.A. Afromowitz, J. Appl. Phys. **44**, 1292 (1973).

49. S.Y. Zhang, J.P. Roger, D. Fournier, A.C. Boccara and Z.C. Wang, Thin Solid films, **186**, 361 (1990).

50. J.P. Roger, F. Lepoutre, D. Fournier and A.C.Boccara, Thin Solid Films, **155**, 165 (1987).

51. L.J. Inglehart, F. Lepoutre and F. Charbonnier, J. Appl. Phys., **59**, 234 (1986).

52. J.W. Shelton and J.A. Armstrong, IEEE J. Quant. Electron. **QE-3**, 302 (1967).

53. H.N. Lin, R.J. Stoner and H.J. Maris, J. Non-destructive Eval., **9**, 239 (1990).

54. C. Thomsen, H.T. Grahn, H.J. Maris and J. Tauc, Phys. Rev., **B34**, 4129 (1986).

55. H.T. Grahn, H.J. Maris, J. Tauc and A. Abeles, Phys. Rev. **B38**, 6066 (1988).

56. A. Harata, H. Nishimura and T. Sawada, Appl. Phys. Lett., **57**, 132 (1990).

PHOTOTHERMAL RADIOMETRY FOR NDE

R. E. Imhof, B. Zhang and D. J. S. Birch

Department of Physics and Applied Physics
Strathclyde University
Glasgow G4 0NG, Scotland

I.	INTRODUCTION	186
	1. Scope and Aims	186
	2. Why Radiometric Detection?	186
	3. Notation	187
	4. PTR Fundamentals	187
	5. Related Literature	189
II.	THEORETICAL BACKGROUND	190
	1. Common Assumptions	190
	2. Calculation Examples	190
	3. Fundamental Scale Lengths	199
	4. Other Relevant Theory	200
III.	APPLICATIONS	202
	1. Measurement of Thermal Properties	202
	2. Measurement of Optical Properties	207
	3. Measurement of Thickness	211
	4. Measurement of Thermal Resistance	213
	5. Material Characterization	214
	6. Flaw Detection and Characterization	218
	7. PTR NDE Measurement Systems	221
	8. Commercial PTR NDE Instrumentation	226
IV.	CONCLUSIONS	227
V.	REFERENCES	227

I. INTRODUCTION

1. Scope and Aims

This chapter focuses on a subset of the vast field of *Photoacoustic and Photothermal Phenomena* that has been or could be applied to Non-destructive Evaluation (NDE) of condensed materials. It focuses on one specific method of measurement, namely infrared radiometry: the detection of photothermal phenomena through their effects on the thermal emissions. Further restrictions of scope, namely the exclusion of work related to semiconductors and photothermal imaging (thermography) have also been observed, because these are subjects of Chaps. 2 and 5 in this Volume. What is left is a research field of astonishing breadth, spanning high precision measurement of fundamental physical constants at one extreme and vague characterizations at the other. Whilst it is true that most published literature is either fundamental or speculative in nature, a growing number of genuine NDE applications is now emerging.

The structure of this chapter is coarsely progressive, from fundamental to applied, from general to particular. The essential unity of the field makes some overlaps and excursions outside the strict domain of the title necessary and desirable, especially to clarify the relationships between radiometric NDE results and the underlying thermal wave physics. By emphasizing these aspects, we hope to aid understanding of what the techniques are capable of and what their limitations are.

2. Why Radiometric Detection?

Like other photoacoustic and photothermal applications, NDE using Photothermal Radiometry (PTR) boils down to the study of non-stationary temperature fields, often described in terms of thermal waves. Thermal wave effects can be detected in a variety of ways: gas microphone cell, piezoelectric, pyroelectric or other transducer in contact with the sample, mirage sensing in the gas above the sample, optical probe beam deflection or diffraction from the surface of the sample, as well as thermal infrared radiometry (see Ref. [1] for an overview). So why single out radiometric detection for separate review?

The answer lies in the special properties of radiometric detection for NDE. For practical NDE, one would ideally look for techniques that can measure without contact, that can be directed remotely to points of interest, that can test samples of arbitrary size and shape, that are capable of on-line monitoring in industrial environments, that are insensitive to noise, vibrations, misalignment, geometric variability and daylight, that require no maintenance or calibration and that cost nothing. The ideal does not exist, but radiometric detection can come close in all but the last criterion.

Another reason why this chapter focuses on radiometric detection is that it adds unique new variables to the measurement. These are associated with the optical properties of the material in the wavelength region of the detected thermal infrared radiation, typically the 6-13µm mid-infrared band for room temperature samples. Although this can make signal interpretation more difficult in some instances, it can also be exploited for NDE.

3. Notation

Two abbreviations are used throughout this chapter, namely NDE for non-destructive evaluation and PTR for photothermal radiometry. The former needs no introduction, but the latter is not universally recognized. In this chapter, we use PTR as a generic term to include all techniques that use optical excitation and radiometric detection, e.g. CMPTR, FIRST, IPPR, OTIS, OTTER, PR, PPTR, RPPR, TIRES, TPTR, TRIR, etc.

Units of measurement and mathematical notation are not uniform within the literature reviewed. In this chapter, we use *SI* units, and the following main symbols:-

α	=	optical absorbance for excitation, ie $\alpha=\alpha(\lambda_{ex})$
β	=	optical absorbance for thermal emission, ie $\beta=\beta(\lambda_{em})$
C	=	specific heat/unit mass
c	=	velocity of light
D	=	thermal diffusivity, ie $D=k/C\rho$
ε	=	thermal effusivity, ie $\varepsilon=(kC\rho)^{1/2}$
h	=	Planck's constant
κ	=	Boltzmann's constant
k	=	thermal conductivity
λ_{ex}	=	wavelength of excitation
λ_{em}	=	wavelength of thermal emission
ρ	=	density of the sample
σ	=	Stefan's constant
θ	=	changes in temperature of the sample
T_0	=	initial temperature of sample
T	=	absolute temperature of sample, i.e. $T=T_0+\theta$.

Other symbols are defined as they arise.

4. PTR Fundamentals

PTR measurement systems perform two separate but closely linked functions. The first is optical excitation, which produces time-dependent and localized temperature fields, often called thermal waves, within the specimen. The second is radiometric detection of the associated time-dependent and localized thermal emissions.

Time-dependent excitation can take many forms, the two extremes

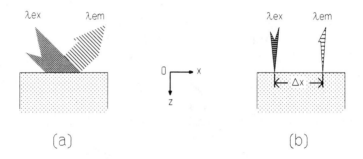
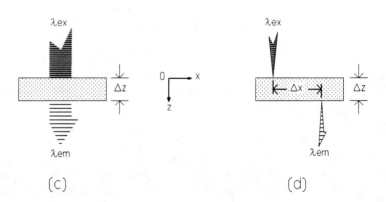

Fig. 1 PTR measurement geometry and co-ordinate system definitions.

being *sinusoidal* modulation and *impulse* excitation. The two are associated with different measurement techniques. The first is a narrow-band measurement of amplitude and phase of a signal at the same frequency as the excitation. The second is a wide-band measurement of the impulse response signal. However, there is a relationship between the two, via the Fourier transformation, if the response of the system is linear.

The other key aspect of such measurements, namely localization, is related to the measurement geometry and focussing. Four measurement geometries used in PTR, and their associated co-ordinate and parameter definitions are illustrated in Fig. 1. Figs. 1(a) and 1(b) illustrate *backward emission* geometries, where excitation and detection occur on the same side of the sample; Figs. 1(c) and 1(d) illustrate *forward emission* geometries, for use with slab samples. In each case, the excitation plane is taken as the $z=0$ plane, with z increasing into the specimen. Slab thickness is denoted by Δz.

The measurements are also affected by focussing or other means of spatially concentrating the radiations. In cases 1(a) and 1(c), relatively large areas of the surface are excited and viewed, so that the effects observed are predominantly from uni-dimensional heat flow, in the z-direction. In cases 1(b) and 1(d), the radiations are focused, so that *transverse heat flow*, in the plane of the sample, cannot be neglected. This offers the possibility of microscopic mapping, and introduces a further measurement variable in the displacement, Δx, that can be introduced between the excitation and emission axes.

Other forms of modulation and measurement geometry have been used or contemplated. Of particular importance to NDE is flying spot scanning, which combines time-dependent excitation with localization and spatial mapping.

5. Related Literature

Many review and overview articles have been published in recent times, which cover various aspects of PTR. In Volume 1 of this Series [2], Boccara and Fournier [3] give an excellent historical perspective of the entire field of photoacoustic and photothermal phenomena. In the same Volume, Busse and Walther [4] review NDE methods based on optically generated thermal waves, and Munidasa and Mandelis [5] review photothermal imaging and microscopy. Photoacoustic sensing techniques and their applications, including radiometric detection and NDE applications, were reviewed by Tam [1]. More specialized overview papers worth consulting include aspects of flash thermal diffusivity measurement [6, 7] and PTR NDE [8, 9, 10(in German), 11, 12(in French), 13 and 14]. In addition, the conference series [15-18] provides a comprehensive overview of the development of the subject within the context of related research.

II. THEORETICAL BACKGROUND

This section presents a theoretical background to PTR in the form of example calculations of a simple but fundamentally important case, the semi-infinite, homogeneous sample, with uniform irradiation, as in Fig. 1(a). Not only does this introduce the main concepts and bring out the main features of what can reasonably be measured by such techniques, but it provides a framework of understanding for more complex cases.

1. Common Assumptions

The theoretical analysis assumes, not unreasonably, that the measurements do not damage or otherwise alter the properties of the sample. The discussion is restricted to solids, whose optical and thermal properties are unaffected by the temperature excursions experienced during a measurement. The excitation is assumed to produce quasi instantaneous heating of the sample through linear absorption. The sample is assumed to be isolated, with negligible heat losses to the surroundings. The contributions to the heat transport from thermal radiation and stress waves within the sample are neglected.

These assumptions exclude semiconductors, materials undergoing dynamic phase transitions and materials which strongly scatter the excitation or thermal emission radiations. They are, however, reasonable for common materials in air, subjected to small temperature excursions near room temperature.

2. Calculation Examples

Two separate steps are required to calculate PTR signals in terms of the physical parameters of the sample and measurement system, namely calculation of localized and time-dependent temperature fields, and calculation of associated changes of thermal emission.

2.1 Temperature Fields

Calculations of time-dependent temperature fields in homogeneous materials resulting from interactions of the optical excitation with the sample start with the thermal diffusion equation

$$\nabla^2 \theta(r,t) = \frac{1}{D} \frac{\partial \theta(r,t)}{\partial t} - \frac{Q(r,t)}{k} \qquad (1)$$

where r is the space co-ordinate (x,y,z) and Q the optically generated heat source. Techniques of analytical solution of Eq. (1) include the separation of variables and the method of images for reducing the dimensionality and simplifying the equations. These, together with Laplace, Fourier and Green's function methods of solution are expounded in the seminal work of Carslaw and Jaeger [19].

For the case of a semi-infinite, homogeneous medium, irradiated uniformly by optical excitation, as illustrated in Fig. 1(a), Eq. (1) reduces to one spacial dimension

$$\frac{\partial^2 \theta(z,t)}{\partial z^2} = \frac{1}{D}\frac{\partial \theta(z,t)}{\partial t} - \frac{I_0 \alpha}{k} e^{-\alpha z} f(t) \qquad (2)$$

where I_0 is the average excitation fluence, and the function $f(t)$ describes its temporal modulation. The boundary condition at the surface is

$$k\left(\frac{\partial \theta(z,t)}{\partial z}\right)_{z=0} = 0 \qquad (3)$$

for no heat transfer to the surrounding air. For such problems, the Green's function for $t>t'$ is

$$G(z,z',t,t') = \frac{e^{-\frac{(z-z')^2}{4D(t-t')}} + e^{-\frac{(z+z')^2}{4D(t-t')}}}{2\varepsilon\sqrt{\pi(t-t')}} \qquad (4)$$

The one-dimensional temperature field, or plane thermal wave, can then be written as

$$\theta(z,t) = I_0 \int_0^\infty dz' \int_{-\infty}^t G(z,z',t,t') \, \alpha e^{-\alpha z} f(t') \, dt' \qquad (5)$$

It then remains to substitute an expression for the modulation function before the integral of Eq. (5) can be solved. For impulse excitation with an ideally narrow impulse, $f(t)=\delta(t)$, one obtains

$$\theta(z,t) = \frac{\theta_0}{2} e^{t/\tau_\alpha} \left\{ e^{-\alpha z} \, erfc\left[\frac{t/\tau_\alpha - \frac{1}{2}\alpha z}{\sqrt{t/\tau_\alpha}}\right] + e^{\alpha z} \, erfc\left[\frac{t/\tau_\alpha + \frac{1}{2}\alpha z}{\sqrt{t/\tau_\alpha}}\right] \right\} \quad (6)$$

where *erfc=1-erf* is the complementary error function, $\theta_0 = E_0 \alpha / \rho C$ is the initial temperature rise at the surface and E_0 is the energy density absorbed from the excitation pulse. The parameter τ_α is an opto-thermal decay time defined by

$$\tau_\alpha = 1/\alpha^2 D \quad (7)$$

and is a key parameter in such measurements. Step excitation, described formally by the Heavyside function $f(t) = \mathcal{H}(t)$, is a related case, since it is the integral of the impulse function. The resulting temperature field is

$$\theta(z,t) = \frac{I_0}{k\alpha} \left\{ \frac{e^{t/\tau_\alpha - \alpha z}}{2} \, erfc\left[\frac{t/\tau_\alpha - \frac{1}{2}\alpha z}{\sqrt{t/\tau_\alpha}}\right] + \frac{e^{t/\tau_\alpha + \alpha z}}{2} \, erfc\left[\frac{t/\tau_\alpha + \frac{1}{2}\alpha z}{\sqrt{t/\tau_\alpha}}\right] \right.$$
$$\left. + \sqrt{\frac{4t}{\pi \tau_\alpha}} e^{-\frac{\alpha^2 z^2}{4 t/\tau_\alpha}} - \alpha z \, erfc\left[\frac{\alpha z}{2\sqrt{t/\tau_\alpha}}\right] - e^{-\alpha z} \right\} \quad (8)$$

According to this, the temperature at every point within the sample, increases with time without limit. In practice, heat losses to the surroundings place an upper limit to the temperature reached.

The most common form of modulation is sinusoidal, characterized by an angular frequency ω. This case is less straightforward, because the excitation necessarily has a steady component of heat input superimposed on the modulated component. The full modulation function can therefore be written as

$$f(t) = \frac{1}{2} \mathcal{H}(t)[1 + e^{j\omega t}] \quad (9)$$

However, since the main interest in such measurements is in the modulated component, the steady heating is usually neglected or treated separately, in which case the solution for the sinusoidally modulated component converges to

$$\theta(z,t) = \frac{\alpha I_0 \mu_D^2 e^{(j\omega t - \alpha z)}}{2k(\alpha^2 \mu_D^2 - 2j)} \left[\left(\frac{\mu_D \alpha}{1+j} \right) e^{-\frac{z}{\mu_D}(1+j-\mu_D \alpha)} - 1 \right] \quad (10)$$

once the initial switch-on transient has decayed. μ_D is the thermal diffusion length, defined by

$$\mu_D = \sqrt{2D/\omega} \quad (11)$$

and is a key parameter of thermal diffusion theory.

2.2 Thermal Emission

The total radiant emissive power, W_{BB}, of a black body in thermal equilibrium can be calculated from Stefan's law:

$$W_{BB}(T) = \sigma T^4 \quad (12)$$

PTR signals come from transient enhancements of emission resulting from time-dependent temperature fields, as discussed above. In the simplest treatment, the time-dependent thermal emissions are assumed to emanate only from the surface of the sample, and the transient temperature excursions are assumed to be small compared with the equilibrium temperature, T_0, so that they can be calculated using a first order Taylor expansion of Eq. (12)

$$\delta W(t) \approx \left(\frac{\partial W}{\partial T} \right)_{T=T_0} \delta T \approx 4\sigma T_0^3 \, \theta(z=0, t) \quad (13)$$

A further refinement for non-black bodies is to introduce a specimen-dependent emissivity factor into Eqs. (12) and (13).

This treatment is unsatisfactory in that it neglects both the spectral distribution of the black body radiation and how this is modified by the spectral properties of the sample. The first is given by Planck's universal function, expressed here in terms of emission wavelength, λ_{em}

$$W_{BB}(\lambda_{em}, T) = \frac{2\pi h c^2}{\lambda_{em}^5 [\exp(hc/\lambda_{em} kT) - 1]} \quad (14)$$

where $W_{BB}(\lambda_{em}, T)$ is the spectral emissive power. The sample dependence can be derived from Kirchhoff's law relating the emittance of non-black bodies

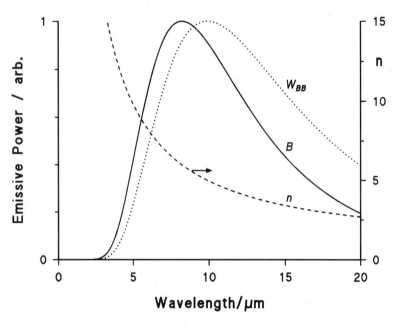

Fig. 2 Blackbody thermal emission at room temperature, T_0=293K. The dotted line (W_{BB}) is the steady emission component, the solid line (B) the transient enhancement. The index n (dashed line) can be interpreted as the percentage increase in emission resulting from a 1% increase in T_0.

to their absorbance, taking into account the arguments of Baltes [20] regarding its application to non-equilibrium emissions. Using again a first order approximation to calculate the time-dependent component of the emissive power from Eq. (14) and restricting the derivation to the case of a homogeneous semi-infinite sample, we obtain

$$\delta W(\lambda_{em}, t) \approx B \int_0^\infty \theta(z,t) \beta e^{-\beta z} \, dz$$

(15)

with $\quad B = \dfrac{n W_{BB}(\lambda_{em}, T_0)}{T_0}, \quad n = \dfrac{\gamma e^\gamma}{(e^\gamma - 1)} \quad$ and $\quad \gamma = \dfrac{hc}{\lambda_{em} k T_0}$

The derivation assumes that the transient thermal emission itself contributes negligibly to the heat transport within the sample.

The main properties of transient thermal emission can be deduced from Eq. (15). The term B in front of the integral describes the black body aspects of the emission, as illustrated in Fig. 2. Note that the peak of the transient emission occurs at shorter wavelengths than that of the steady black body spectrum, because it is related to the first derivative of $W_{BB}(\lambda_{em},T)$. The function n can be interpreted as the percentage increase in emission resulting from a 1% increase of absolute temperature [21]. Its increase towards shorter wavelengths indicates an increasing ratio of transient to background emission. The integral of Eq. 15 takes into account the spectral properties of the sample in the region of thermal emission. It tends to zero in the limit $\beta \to 0$, as expected for a transparent body. In the other limit, $\beta \to \infty$, it tends towards $\theta(z=0)$, to give the same radiation as a black body of temperature equal to the surface temperature. If β is finite but independent of wavelength, then the integral evaluates to the constant emissivity of a grey body. However, many materials of interest in NDE have structured absorption spectra in the wavelength range of interest, and Eq. (15) provides the basis for studying these, as reviewed further in Section III(2.2).

2.3 PTR Impulse Response

The form of any PTR signal can now be calculated. For example, consider the impulse excitation of a semi-infinite, homogeneous material, whose temperature field is given in Eq. (6). The signal, $S(t)$, integrated according to Eq. (15) is given by

$$S(t) = \frac{\zeta E_0 \alpha \beta}{C \rho (\beta^2 - \alpha^2)} \left\{ \beta e^{t/\tau_\alpha} \mathrm{erfc}\left[\sqrt{t/\tau_\alpha}\right] - \alpha e^{t/\tau_\beta} \mathrm{erfc}\left[\sqrt{t/\tau_\beta}\right] \right\} \quad (16)$$

where the parameter $\zeta = \zeta(\lambda_{em})$ includes factors that depend on the black body emission curve, detector sensitivity, focussing and alignment, but are independent of the properties of the sample *per se*. The decay time $\tau_\beta = 1/\beta^2 D$ is defined in an analogous way to Eq. (7).

Equation (16) has the form of an instantaneous signal jump at time $t=0$, followed by a decay which is rapid at first, but becomes increasingly languid with time. Its properties are worth exploring in detail, because our understanding of more complex cases can be based on deviations from this most fundamental of meaningful cases. The impulse response, rather than the more common sinusoidal modulation was chosen for this example, because it is less thoroughly discussed in the literature. However, it should be straightforward to interpret the points made, since, in this linear theory, the impulse and sinusoidal modulation signals are related by the Fourier transformation. Equivalent step excitation results can be calculated from indefinite integrals with respect to time of the impulse solutions. The following properties of Eq. (16) can be deduced:-

i) It is symmetrical with respect to interchange of the optical absorbances α and β. This is a consequence of a more general reciprocity property, explored by Kuo et al for the case of photoacoustic gas cells [22, 23]. It can be demonstrated for the case of radiometric detection from the symmetry properties of the Green's function (see also Ref. [5]).

ii) The limit $t \to 0$ of Eq. (16) can be expanded as a power series

$$\lim_{t \to 0} S(t) \approx \frac{\zeta E_0 \alpha \beta}{C\rho(\alpha + \beta)} \left[1 - \alpha\beta D t + \frac{4\alpha\beta(\alpha + \beta)}{3\sqrt{\pi}}(Dt)^{3/2} - \cdots \right] \quad (17)$$

This shows that the initial amplitude, $S(0)$ does not depend on the thermal conduction properties of the material, but only upon its optical properties and the volumetric specific heat. We call this the *optical limit*, a unique feature of impulse excitation. The initial gradient of the signal, normalized to unity initial amplitude, is

$$\lim_{t \to 0} \left(\frac{1}{S(0)} \frac{\partial S}{\partial t} \right) \approx -\alpha\beta D \quad (18)$$

and is an *opto-thermal* quantity, because it depends inextricably on optical and thermal properties of the sample.

iii) The limit $t \to \infty$ of Eq. (16) can also be expressed as a power series

$$\lim_{t \to \infty} S(t) \approx \frac{\zeta E_0}{\varepsilon\sqrt{\pi t}} \left[1 - \left(\frac{\alpha^2 + \beta^2}{\alpha^2 \beta^2} \right) \frac{1}{2Dt} + \cdots \right] \quad (19)$$

This shows that the decay at long times becomes increasingly languid, until the limiting $t^{-1/2}$ law of uni-dimensional diffusion is obtained. In this limit, the signal does not depend on the optical properties of the material, but is inversely proportional to its effusivity. We call this the *thermal limit* of the impulse response curve.

iv) The limit $\beta \to \infty$ whilst α remains finite is equivalent to the limit $\alpha \to \infty$ whilst β remains finite. Denoting the finite absorbance, whether α or β, by $\hat{\alpha}$ and the corresponding opto-thermal decay time by $\hat{t} = \hat{\alpha}^{-2} D^{-1}$, Eq. (16) reduces to

$$S(t) \approx \frac{\zeta E_0 \hat{\alpha}}{C\rho} \exp(t/\hat{\tau})\, erfc\left[\sqrt{t/\hat{\tau}}\right] \tag{20}$$

v) The limit $\alpha = \beta$ leaves Eq. (16) undefined. This has no physical relevance, although it does present practical problems for data analysis. The signal in this *symmetrical limit* can be calculated using Ref. [24]

$$S(t) = \frac{\zeta E_0 \hat{\alpha}}{2C\rho}\left\{\left(1 - \frac{2t}{\hat{\tau}}\right)\exp(t/\hat{\tau})\, erfc\left[\sqrt{t/\hat{\tau}}\right] + 2\sqrt{\frac{t}{\pi\hat{\tau}}}\right\} \tag{21}$$

Fig. 3 Semi-opaque and symmetrical limits of the impulse response of a semi-infinite homogeneous material, as in Eqs. (20) and (21). The general solution of Eq. (16) describes a family of similar curves within these two extremes.

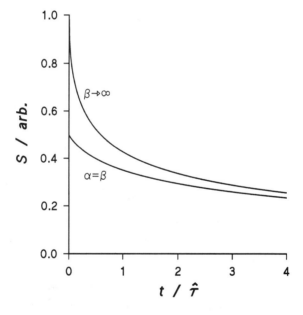

The two limiting forms, Eqs. (20) and (21), are illustrated in Fig. 3. They represent the extremes delimiting the general form of Eq. (16). Their main differences are in the initial slopes and curvatures in the opto-thermal region, $t/\hat{t} < 3$. The initial amplitude of Eq. (21) is half that of Eq. (20), because of the loss of sensitivity implied by the finite second absorbance. As t increases, the two curves converge, rapidly at first, to differ by less than 10% at $t/\hat{t} = 3$. As $t \to \infty$, they tend towards the $t^{-1/2}$ decay law of the thermal limit. However, an optical memory is retained for a long time because the difference between the two curves tends to the slowly varying function $\hat{t}/2t$.

2.4 Impulse Response Data Analysis

Impulse PTR measurements usually use digital transient capture, with signal averaging over several excitation pulses where necessary, to record impulse response curves over a time interval from before excitation to some time after excitation, as illustrated for example in Fig. 6. Such impulse response curves are useful in their own right, because they contain a wealth of information about the sample. They can readily be compared visually with similar curves in a compatible measurement series, to give a qualitative appreciation of properties or phenomena under study. To do so, it is important to pre-process the raw data in the following ways, in order to bring out their salient features:

i) Background signal levels need to be subtracted, since these represent instrumental offsets of no physical significance. Background levels can be calculated from data points prior to excitation.

ii) For comparing impulse response shapes and decay rates, the peak signal amplitudes can usefully be normalized.

iii) For displaying certain aspects of the impulse response curves, it can sometimes be useful to use logarithmic or otherwise non-linear scales for amplitude and/or time axes.

It is often important to reduce the volume of data to but a few physical parameters. This can sometimes be done by focussing on one specific aspect of a family of impulse response curves, such as initial amplitude for optical parameter measurement, or late decay data for thermal effusivity measurement. The measured impulse response curves can also be transformed, using a Fourier transformation for evaluation in the frequency domain, for example. In cases where a theoretical model is available, the entire impulse response measurement can be evaluated using least squares curve fitting. The semi-infinite, homogeneous model, Eq. (16), for example, can reduce a measured impulse response curve to just three parameters: an initial amplitude and two opto-thermal decay times.

$$S(t) = \frac{\zeta E_0}{\varepsilon (\tau_\alpha - \tau_\beta)} \left\{ \sqrt{\tau_\alpha}\, e^{t/\tau_\alpha} erfc\left[\sqrt{t/\tau_\alpha}\right] - \sqrt{\tau_\beta}\, e^{t/\tau_\beta} erfc\left[\sqrt{t/\tau_\beta}\right] \right\} \quad (22)$$

Of these, the initial amplitude is a relative parameter, depending on the energy absorbed, focussing, alignment and other instrumental factors, whereas the decay times are absolute parameters, depending only upon the properties of the sample. It is the availability of such absolute parameters in a single, rapid measurement that makes impulse PTR so attractive for NDE.

The full model, as expressed in Eqs. (16) or (22), is difficult to use in practice, because:

(a) If τ_α and τ_β are incommensurate, then the shorter decay time contributes little to the model, and the limiting form of Eq. (20) needs to be used. This is a manifestation of photothermal saturation [25]. The fitted opto-thermal decay time, $\hat{t}$, in this case relates to the smaller of the two absorbances, α or β.

(b) If $\tau_\alpha \approx \tau_\beta$, then Eq. (21) needs to be used, because of cancellation inherent in Eq. (16). The fitted opto-thermal decay time, $\hat{t}$, in this case can be interpreted as an average of two similar values.

Of course, least squares curve fitting can also be useful when no *ab initio* theories are available. In such cases, empirical functions can be used for characterization and data reduction.

3. Fundamental Scale Lengths

The descriptions of PTR signals and their information content can be made conceptually more appealing by framing them in terms of fundamental scale lengths. The absorbances α and β in Eq. (16), for example, can be expressed as absorption lengths,

$$\mu_\alpha = \frac{1}{\alpha} \quad \text{and} \quad \mu_\beta = \frac{1}{\beta} \quad (23)$$

equal to the characteristic $1/e$ penetration distances of the radiation into the material.

The other fundamental scale length of thermal diffusion theory is thermal diffusion length, μ_D. This has already been introduced in Eq. (11), in association with sinusoidally modulated excitation. Impulse and step excitation, however, have no unique diffusion length, because the heat continues to spread as time proceeds. Instead, a time-dependent diffusion length for uni-dimensional problems can be defined by a relationship such as

$$\mu_D \approx \sqrt{2Dt} \qquad (24)$$

which comes from the variance term of the Green's function, Eq. (4), and is equal to the RMS spreading of heat occurring in time t. This definition is not unique, but other definitions would deviate from it only by a numerical factor of the order of unity.

Eq. (16) can now be analyzed in terms of the three scale lengths, μ_α, μ_β and μ_D:

i) In the optical limit at $t=0$, the thermal diffusion length, $\mu_D=0$ and $S(t \to 0)$ is independent of μ_D and inversely proportional to the sum, $\mu_\alpha+\mu_\beta$.

ii) In the thermal limit, $t \to \infty$ and $S(t \to \infty)$ is independent of μ_α and μ_β and inversely proportional to μ_D. At intermediate times the signal has opto-thermal character and therefore depends on all three scale lengths. The strongest influence is exerted by the largest of them, as is often the case in other contexts.

The impulse response function, Eqs. (16) or (22), has two opto-thermal decay times, τ_α and τ_β, as parameters. From Equations (7), (23) and (24) these are characteristic times when the corresponding absorption lengths are commensurate with the thermal diffusion length. At shorter times, the thickness of material probed is determined mainly by the larger of the two optical absorption lengths. At longer times, the depth probed increases as the thermal diffusion length increases. This is the basis of *depth profiling*, which gives PTR the ability to sense changes in material properties with depth. The relationship between time and depth probed gives PTR impulse response curves a somewhat analogous meaning to A-scans in ultrasonic NDT. Depth profiling can also be performed with sinusoidal excitation, where the thermal diffusion length is determined by the modulation frequency, as in Eq. (11).

4. Other Relevant Theory

The above conclusions were reached for a semi-infinite, homogeneous sample with one-dimensional heat flow. In practice, they apply to samples of quite modest dimensions, because boundaries that are distant in terms of the fundamental scale lengths can be ignored. These scale lengths also determine whether the optical and thermal wavefronts can be regarded as plane. However, most interest in practical NDE is in materials that are inhomogeneous in one or more ways. Theoretical understanding of the relationships between sample properties and PTR signal is an essential prerequisite to measurement, but the mathematics involved quickly becomes unmanageable.

Some conceptual simplifications are possible, however. Heat flow in layered materials, for example, can be understood by modelling the effect of boundaries between materials in terms of reflection and transmission. For the simplest case of plane thermal waves impinging normally on a plane boundary [26], the coefficients of reflection, R, and transmission, T, are given by

$$R = \frac{1 - \varepsilon_2/\varepsilon_1}{1 + \varepsilon_2/\varepsilon_1} \qquad T = \frac{2}{1 + \varepsilon_2/\varepsilon_1} \qquad (25)$$

where subscripts 1 and 2 represent the source and destination media respectively. Note that these coefficients are non-dispersive and depend only on media properties, not thermal-wave frequencies. Also, the effect of an interface depends on the adjoining effusivities, not diffusivities. Thus, an interface between alumina ceramic and stainless steel, for example, would produce little perturbation to thermal wave propagation, despite a thermal diffusivity ratio of 2.4:1, because the effusivity ratio is close to unity. Finally, the resultant temperature at any point can be calculated by superposing all reflected and transmitted thermal wave components at that point. Interference effects [14,26] can therefore be observed, although these tend to be weak, because of the highly damped nature of thermal waves.

In many cases, inhomogeneities have characteristic lengths, such as layer thickness, mean length and separation between reinforcing fibers, mean diameter and separation between particles, etc, associated with them. In addition, it may sometimes be of advantage to introduce controlled inhomogeneities, such as the diameter of the focused excitation spot, or the separation between excitation and emission spots, into the measurement system, in order to achieve a given measurement purpose. In all such cases, the resulting PTR signals may be interpreted in terms of the interplay between these characteristic lengths and the fundamental scale lengths introduced in Section II(3).

The above conceptual crutches apart, the modelling of PTR measurements for inhomogeneous materials and measurement systems frequently is too complex for general treatment. The modelling of temperature fields *per se* is more developed than the optical aspect, because such models have much wider application than just PTR (see Volume I of this Series, for example). In any case, the thermal limit is an appropriate approximation in many PTR experiments. Work on temperature fields of particular interest for PTR NDE include laser heating calculations for stationary [27] and moving beams [28], a formulation of the inverse problem of determining thermal conductivity profiles from surface thermal wave measurements by Vidberg *et al* [29], the thermal diffusivity inverse problem by Mandelis *et al* [30-33]; and Aamodt *et al*'s analysis of thermal transit times in multi-layered coatings [34].

When optical absorption, reflection and scattering are included, the problems quickly acquire too many parameters, many of which are unknown and perhaps unknowable. Such problems need to be tackled using approximations, semi-empirical models, numerical simulation, and, if all else fails, empirical characterization. Most theoretical work on the interpretation of PTR measurements is therefore to be found alongside the experimental work reviewed in Section III. Notable exceptions include Tom *et al*'s generalized model of PTR [35]; Aamodt and Murphy's study of materials with continuously varying optical and thermal properties [36], Mandelis *et al*'s study of continuously thermally inhomogeneous materials [30,31]; and the work of Baumann and co-workers on thermal waves in multi-layered media [37-39].

III. APPLICATIONS

The PTR-NDE applications presented in this section are too diverse to yield to orderly sub-division. They are therefore presented in a coarse grid of application areas and techniques along a trend line from fundamental to applied. In the absence of other criteria, a broadly historical approach is used within each section.

Fig. 4 Calculated temperature history curve of flash diffusivity measurement, for a material of thickness Δz and finite optical transparency. The excitation is assumed to be strongly absorbed, with $\alpha \Delta z = 20$. The absorbance for thermal emission varies as follows: (1) $\beta \Delta z = 20$, (2) $\beta \Delta z = 3$, (3) $\beta \Delta z = 1$, (4) $\beta \Delta z = 0.3$. (From Ref. [44], with permission).

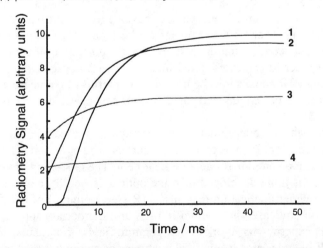

1. Measurement of Thermal Properties

A knowledge of thermal properties is fundamental to all aspects of material testing and a cornerstone of PTR NDE techniques.

1.1 Forward Emission

The first reported PTR technique was the flash diffusivity technique of Rudkin *et al* in 1962 [40]. It used a forward emission geometry, as illustrated schematically in Fig. 1(c), to measure thermal diffusivities of slab samples. Impulse excitation from a Xe flash was used to produce rapid and uniform heating of one face of the sample, typically a metal disk of thickness of the order of millimeters. The time-dependent temperature of the rear face was measured by means of a PbS detector and recorded on Polaroid film from an oscilloscope display. The form of such signals for the opaque materials studied is illustrated in Fig. 4, curve (1). For this special case, the relationship between thermal diffusivity, sample thickness, Δz, and $\Delta t_{1/2}$, the time taken for the temperature of the rear face to rise to one half its final value above ambient, has the simple form

$$D = 0.139 \frac{\Delta z^2}{\Delta t_{1/2}} \tag{26}$$

Deem and Wood [41] used a Ruby laser in place of the Xe flash in an otherwise similar experiment. Taylor [42] showed that the method could be used for two-layer samples, with measurements on a 0.75mm thick stainless steel substrate carrying a 60µm painted layer. Korshunov *et al* [43] used sinusoidally modulated excitation to measure thermal diffusivities of two-layer metallic systems.

Thermal diffusivity measurements of polymer films are more difficult, because of their finite transparency to both excitation and thermal emission radiations. Choy *et al* [44] adapted the forward emission flash technique for polymer studies by using 266nm excitation radiation from a frequency quadrupled Nd:YAG laser, which is strongly absorbed by most polymers. Transparency effects are nevertheless evident in the measured response curves, as illustrated in Fig. 4. This is partly because samples as thin as 14µm were studied and partly because the absorption spectra of many polymers have regions of low absorbance in the 8-13µm wavelength range used for detection. They derived theoretical expressions that took optical absorbance effects into account in the least squares analysis of the measured response curves. A different approach was adopted by Tsutsumi and co-workers [45, 46] in that they used thin carbon coatings on one or both faces of the samples to create essentially opaque faces for measurement. Frederikse *et al* [47] reported the measurement of thermal diffusivity of an isolated CVD diamond

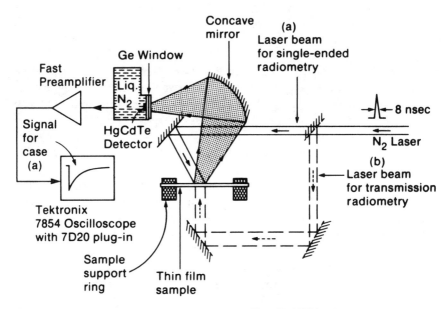

Fig. 5 PTR apparatus for thin film measurements. (From Ref. [50], with permission).

film, 0.25mm thick, by means of a sinusoidally modulated PTR technique, again using coatings to eliminate optical transparency effects.

1.2 Backward Emission

The forward emission techniques above require sample preparation and access to front and back faces, which restrict their usefulness for NDE. The method developed by Schultz [48] overcomes these restrictions by using a backward emission geometry, as in Fig. 1(a). It used steady and uniform heating of the surface with radiant energy, suddenly switched on, whilst observing the characteristically shaped surface temperature rise using an infrared radiometer. With the highly opaque coated graphite samples used, the surface temperature rise is approximately given by

$$\theta(0,t) = \frac{2I_0}{\varepsilon}\sqrt{\frac{t}{\pi}} \qquad (27)$$

This equation can be derived from the step response, Eq. (8), as the limit ($z \to 0$, $\alpha \to \infty$, $\beta \to \infty$). Thermal inertia values ($= \varepsilon^2$) were deduced by comparing the measured response curves with suitable reference standards. Cielo et al [49] also used step excitation for measuring thermal effusivities of porous ceramic materials, but incorporated a reflective cavity in contact with the sample, similar to that illustrated in Fig. 14a, to eliminate surface emissivity as an unknown quantity.

Leung and Tam [50] showed that impulse excitation could be used in a backward emission arrangement to measure thermal diffusivity or thickness of isolated thin films, by exploiting the effect of the finite thickness on their impulse response curves. A schematic diagram of the apparatus is shown in Fig. 5. The impulse response of an isolated thin film resembles that of a semi-infinite sample, i.e. Eq. (16), as long as the fundamental optical and thermal scale lengths are small compared with its thickness. However, thermal diffusion is hindered, once the thermal diffusion length becomes comparable with the film thickness. The slope of the decay curve then reduces to a low value, determined by thermal diffusion in the plane of the film and heat loss to the surroundings. The technique was used on stainless steel samples of varying thickness, using a fitting technique to extract thermal diffusivities from the measured impulse response curves, as illustrated in Fig. 6. Measurements on a 89μm thick Teflon film were found to require correction

Fig. 6 (a) Impulse response curves observed in backward emission for stainless steel samples of different thickness. (b) Curve fitting example, for thermal diffusivity determination. (From Ref. [50], with permission).

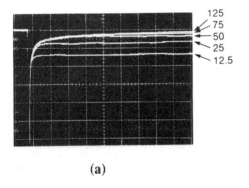

(a)

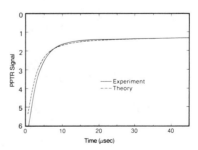

(b)

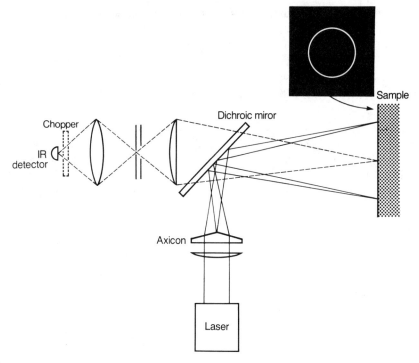

Fig. 7 Optical system for converging thermal wave measurements. (From Ref. [57], with permission).

for optical transparency. See Ref. [51] for a detailed theoretical analysis of this and related aspects of impulse PTR.

The backward emission methods described above can also be used to measure thermal diffusivities of homogeneous, semi-infinite materials. Measured impulse response curves can be analyzed using the theory of Section II, to obtain opto-thermal decay times, from which thermal diffusivities can be calculated, once the optical parameters are known. However, apart from some notable exceptions, [52, 53], this procedure is not useful in practice, because of uncertainties associated with the optical constants at both excitation and thermal emission wavelengths.

Frederikse and Ying [54] used a sinusoidally modulated CO_2 laser to measure thermal diffusivities of thin oxide coatings on stainless steel by a comparative technique, using reference samples whose thickness was large compared with the thermal diffusion length.

1.3 Transverse Heat Flow

The techniques reviewed thus far all used excitation over relatively large areas of the surface, so that transverse heat flow can be neglected. If focussing elements are used for excitation and thermal emission, then a large variety of 3-dimensional heat flow patterns can be generated, as indicated schematically in Figs. 1(b) and 1(d). Beyfuss and co-workers [39, 55, 56] used this approach to measure thermal parameters of layered samples. They measured amplitude and phase of the infrared emission at the center of a focused argon ion laser, modulated over a range of frequencies. By comparing the results with calculations of 3-dimensional heat flow, they were able to measure thermal conductivities and heat capacities of thin films of interest in microelectronics.

A different method of producing transverse heat flow was pioneered by Cielo *et al* [57]. It used an axicon, an optical element of conical symmetry, to form a ring of excitation on the sample surface, with the infrared detector focused to view a small spot at its centre, as illustrated in Fig. 7. This measurement geometry is analogous to that in Fig. 1(b), but with the excitation spread into a circle of radius Δx, to produce rotational symmetry about the emission optical axis. It has the advantage of high sensitivity for a given excitation fluence. Enguehard *et al* [58, 59] measured thermal radial diffusivities of anisotropic materials, especially laminates, using this technique with pulsed ruby laser excitation. Other geometries were discussed by Cielo *et al* [60] in the context of optically generated acoustic waves, rather than PTR.

Frederikse *et al* [47] measured thermal diffusivities of very thin CVD diamond films deposited on insulating substrates by an interesting variant of the transverse heat flow technique. A sinusoidally modulated laser beam was used to heat a blackened spot on the film, while the temperature of another blackened spot, a distance of 1-2mm from the heated spot, was measured radiometrically. The thermal diffusivity was calculated from the dependence of the signal on modulation frequency, with allowance made for heat losses to the substrate.

2. Measurement of Optical Properties

The optical measurement capabilities of PTR come from two distinct aspects of the technique, associated, respectively with the distribution of optical excitation and enhanced thermal infrared emission radiations within the sample. They can be used to measure excitation and thermal emission spectra respectively, as described below.

2.1 Excitation Spectroscopy

PTR can be used to measure excitation spectra of essentially *opaque* materials in a directly analogous way to photoacoustic spectroscopy, as first pointed out by Kanstad and Nordal [61]. Indeed, the desire to overcome the limitations of photoacoustic spectroscopy in its need for an enclosed sample volume, sensitivity to extraneous noise and limited speed of response is at the

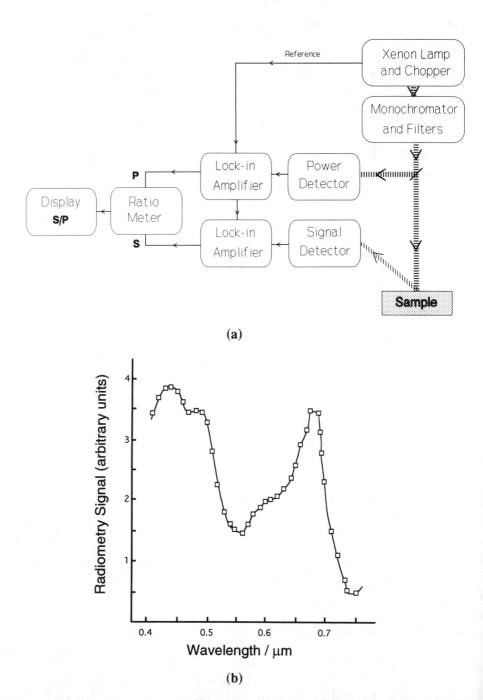

Fig. 8 (a) Apparatus for measuring PTR excitation spectra. (b) PTR excitation spectrum of ivy leaf [62].

roots of this discovery. They used a backward emission geometry, chopped excitation and ratiometric signal processing with lock-in amplification, to correct for fluctuations in incident power and reduce noise. The apparatus, illustrated in Fig. 8(a), was used to measure visible and infrared excitation spectra of powders and surfaces (eg Fig. 8(b)), as described in a series of important papers [25, 62-65], in which photothermal saturation criteria and detection sensitivity for both sinusoidal modulation and impulse excitation were also discussed for various classes of samples. Tam and Sullivan [66] showed the first excitation spectrum measured using impulse excitation.

2.2 Emission Spectroscopy

Imhof *et al* [67, 68] first showed that PTR transient emissions carry spectral information. They measured a crude transient thermal emission spectrum of TiO_2 by interposing a series of interference filters between the sample and infrared detector in an otherwise conventional impulse PTR set-up. Whilst the $\approx 10\%$ spectral resolution of such interference filters is too poor for meaningful infrared spectroscopy, the technique nevertheless provides useful information and an element of control over the conditions of measurement. The latter point is illustrated in Fig. 9 for a *VISA* card of laminated plastic [13]. The experiment used 532nm excitation, which was only weakly absorbed by the clear top layer of the card, but strongly absorbed by the blue coloured plastic below. The curves were measured under identical conditions, except for the thermal emission wavelength. The front curve, at $\lambda_{em}=12.0\mu m$ looks through the top layer, at emissions from the lower layer. The curves behind used thermal emission filters of progressively shorter wavelengths. The rearmost curve, at $\lambda_{em}=7.8\mu m$, looks at emissions from the

Fig. 9 PTR emission spectrum of a laminated plastic. From front to rear, the thermal emission wavelengths are 12.0, 11.5, 10.2, 9.5, 9.1, 8.5 and 7.8µm [13].

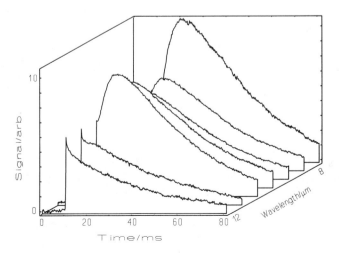

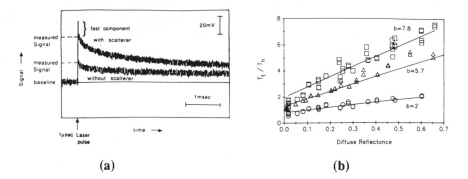

Fig. 10 (a) PTR decay curves for gels with and without added scattering material. This shows how turbidity increases signal strength and causes a rapidly decaying component to appear. (b) Dependence of sub-surface fluence on diffuse reflectance for three different conditions of refractive index change at the surface, as indicated by the parameter b. (From Ref. [71], with permission).

top surface, which are delayed with respect to the excitation pulse, because of thermal diffusion through the top layer. The signals at intermediate wavelengths show different admixtures of decay and delayed thermal wave, which are directly related to changes in $\beta(\lambda_{em})$.

Jones and McClelland exploited the high resolution and throughput of a *Fourier Transform Infrared Spectrometer* to measure transient thermal emission spectra from various plastics and inorganic materials using both cw [69] and impulse [70] laser excitation. Excitation using air jets was also used, but this is not relevant here. The technique shows very considerable promise for NDE, since it offers capabilities for non-invasive, on-line spectral analysis of opaque materials of comparable resolution to that achieved by conventional laboratory analysis.

2.3 Other Optical Measurements

Optical scattering within turbid media can affect the distribution of both excitation and emission radiations. Whilst understanding of such problems is well advanced within other contexts, the work of Anderson *et al* [71] appears to be the only systematic investigation of the effects of scattering on PTR measurements. They used aqueous gels containing absorbing dye, together with various scattering materials, to model the optical properties of human skin. Fig. 10(a) illustrates the difference in PTR signals observed with and without scatterer in otherwise identical samples. Without scatterer present, the decay law was found to follow that of a homogeneous, thermally thick material, Eq. (20). The addition of scattering material was found to increase the signal strength, and to cause a rapidly decaying initial component to appear. Experiments were performed to relate signal strength to diffuse reflectance, used as a measure of turbidity. The results, summarised in Fig. 10(b) showed that sub-surface energy density, taken as proportional to initial temperature jump minus fast component, increased with turbidity, and refractive index difference at the boundary of sample and surroundings. The main conclusion, that scattering could increase sub-surface energy density in comparison with homogeneous media, was thought to have important implications for *in-vivo* phototherapy.

Another optical NDE application of PTR is the measurement of absorption in low loss optical coatings, whether transmitting or reflecting. Conventional absorption measurement techniques cannot reliably distinguish absorption from other losses, such as scattering, and become insensitive as absorption losses decrease. Yet even very low absorption can lead to optical damage in high energy optical components. Impulse PTR gives a direct measurement of absorption, because the signal strength at $t=0$ depends only on energy absorbed and is independent of thermal parameters, reflection or scattering, as illustrated for a set of ZnS/NaF coated interference filters in [72].

3. Measurement of Thickness

Measurements of thickness of slabs or coatings are related to measurements of thermal diffusivity, since both quantities appear in the equations used for data evaluation. However, thickness is the parameter of primary interest in the work reviewed here.

The first PTR measurements of material thickness were reported by Busse [73], who used a sinusoidally modulated argon laser and Golay detector, to measure the thickness of an aluminum wedge. Thickness in excess of 4mm, many times the thermal diffusion length, could be measured in the forward emission geometry used. Measurements using backward emission geometry are of greater interest in NDE, however, especially for measuring coating thickness. Thermal wave reflection at the interface was exploited by Sheard and Somekh [74] for determining the thickness of opaque films independently of the properties of the substrate. They used a sinusoidally modulated diode laser of 860nm wavelength in a novel design of photothermal microscope, illustrated in Fig. 4 of Chap. 5 in this Volume, to measure the phase and amplitude of the thermal radiation signal over a range of frequencies. Busse and co-workers [11, 75, 76] investigated two-layer paint

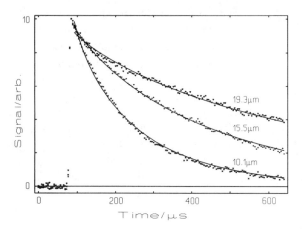

Fig. 11 Measured decay data with best fit curves calculated from Eq. (28) superimposed, for electrocoated steel panels [77].

Fig. 12 (a) Model of a thermal contact, where heat transfer can take place via solid contact, conduction through gas and radiation; (From Ref. [82], with permission). (b) PTR decay curves for a film separated from a heat sink by a variable air gap. (From Ref. [84], with permission).

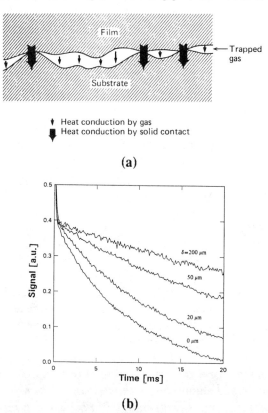

on polymer systems where either or both layers were opaque or where the top layer was transparent to the excitation.

The above work used sinusoidally modulated excitation. Impulse excitation was used by Imhof et al. to measure the thickness of paint on steel [77]. The method relied on the ability of a thermally conducting substrate to conduct heat deposited in a thermally insulating coating away at a rate that increased as coating thickness decreased, as illustrated in Fig. 11. A rigorous general solution to this problem is not available, but the semi-empirical expression,

$$S(t) \propto \exp(t/\hat{t}) \, erfc\left[\sqrt{t/\hat{t}}\right] e^{-t/\tau_c} \qquad (28)$$

was found to have wide validity. It was derived from the limiting forms of the general solution at very short and very long times [78]. In the above expression, the decay time $\hat{t}$ has the same meaning as in Eq. (20) and $\tau_C = (\Delta z)^2/sD$ is a characteristic decay time of the coating, from which thickness or thermal diffusivity can be calculated. The constant s gives a measure of the effectiveness of the substrate as a heat sink. The use of least squares curve fitting gave a high sensitivity to thickness in measurements that were insensitive to instrumental parameters such as energy absorbed, angle of incidence and alignment. A similar approach was used by Cielo and Dallaire [79] with thermal barrier coatings. Imhof et al [72, 80] also developed a technique for measuring the thickness of clearcoats of car paint systems by measuring the transit times of delayed thermal waves, similar to those illustrated in Fig. 9. Tam et al [81] also reported the use of impulse PTR for NDE of layered thin films, including the measurement of thickness and coating weight.

4. Measurement of Thermal Resistance

Thermal resistance is an interfacial property that determines the rate at which heat can be transferred between two adjacent layers that are in less than perfect thermal contact. Figure 12(a) [82] illustrates a model of thermal contact, in which heat transfer is assumed to take place via conduction at areas of solid contact, as well as conduction through gas and radiation in the gaps. A knowledge of thermal resistance values and their dependence on the parameters of the contact between the surfaces and the gas in the gap is clearly a pre-requisite to estimating the size of cracks, disbonds and similar flaws in PTR NDE.

Tam and co-workers [82-84] conducted fundamental investigations of thermal resistance at contact interfaces using impulse PTR in backward emission. A measurement of the effect of an air gap between a thin film and a heat sink is illustrated in Fig. 12(b). Their work resulted in a most useful body of data on thermal resistance dependence on gas type, gas pressure, contact pressure and gap size, that adds considerably to fundamental understanding. Beyfuss *et al* [85] showed that thermal resistance changes between a silicon chip and the heat sink to which it was bonded could be mapped using phase-shift measurement in a backward emission geometry. Excitation was by means of a chopped, focused argon laser beam, scanned in a raster pattern across the surface.

5. Materials Characterization

A great deal of NDE falls into this category. The rigor of designing systems and procedures for measuring specific and fundamental material properties is abandoned in favor of practical considerations: speed and ease of measurement, cost and flexibility. The properties of interest frequently have mixed parentage: microscopic and macroscopic, qualitative and quantitative. For practical NDE, the measurement needs to be sensitive to the property of interest, while remaining insensitive to other variables. This is often quite difficult with PTR, because the signal itself has mixed parentage: optical and thermal. Frequently, such problems are addressed through the use of relative, rather than absolute measurements, using either separate calibration samples, or by mapping out changes within a single sample.

5.1 Films, Coatings and Composites

Metallic, ceramic and polymeric coatings are widely used in industry to produce specific surface characteristics of engineering components, from thermal barriers to corrosion resistance. Their characteristics vary widely, depending on the materials involved and the conditions under which they are produced. Few methods of non-destructive characterization exist, and their applicability in practice is limited. For this reason, considerable attention has been focused on thermal methods, including PTR. Measurement conditions producing short diffusion lengths can be used to characterize the coating itself, provided that the optical absorption lengths are similarly small. Longer diffusion lengths can be used to probe the substrate and its interface with the coating.

Almond and co-workers [86, 87] used thermal wave interference measurements with a backward emission set-up to characterize ceramic coatings. Heuret *et al* [88] were able to measure thermo-physical parameters of plasma sprayed ceramic materials by developing a theoretical model to account for their absorption and scattering properties. The characterization of carbon fiber reinforced plastics using stationary and moving focused laser excitation was reported by Rief *et al* [89]. Tam *et al* [81] constructed and evaluated an impulse PTR apparatus capable of rapid on-line measurements on thin layered films. Capabilities for characterization included coating weight, thickness, composition, moisture content and structural defects.

Results were presented of measurements on printer ribbons.

Another area where characterization of coatings is of value is the monitoring of changes in paint and polymer coatings with age and exposure to hostile environments. Imhof et al [90] showed that opto-thermal decay times measured in backward emission with pulsed UV excitation correlated with weathering damage, especially in the early stages of the process. Later work [77] showed that the sensitivity could be enhanced through selective infrared detection. Impulse response PTR is particularly suited to condition monitoring over extended periods, since opto-thermal decay time is a material parameter whose measurement does not rely on alignment, focussing, excitation energy or other instrumental parameters.

5.2 Drying, Curing and Moisture Content

These topics are presented under the heading of characterization rather than measurement because, although the aims of the experiments themselves are specific, the PTR signals relate to combinations of physical parameters, including density, specific heat, thermal conductivity and optical absorbance, that change in generally unknown ways during the processes monitored.

Krapez et al [91] developed an *in situ* method of testing the degree of cure of carbon fiber reinforced plastics using step excitation in a backward emission set-up. The laser was used to heat the surface to the curing temperature while the infrared emission was monitored to give a measure of surface temperature. Typical results for three heating cycles with a graphite-epoxy prepreg sheet are shown in Fig. 13. The thermal history curve for the first heating cycle shows clear evidence of laser-induced exothermic curing taking place. This was used as a reference against which similar history curves subsequent to curing were compared, to give a measure of degree of

Fig. 13 Surface temperature increase for a graphite-epoxy prepreg sheet. Curves (a), (b) and (c) are for first, second and third heating cycles. (From Ref. [91], with permission).

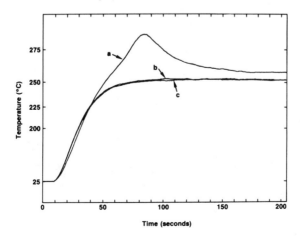

cure that was relatively insensitive to variations in geometry and other material properties. Leung and Tam [92] devised a technique based on impulse PTR in backward emission to measure the thermal conductivity of an epoxy bond beneath an opaque surface layer, from which degree of cure could be derived. Paint drying was investigated by Busse *et al* [75] using sinusoidally modulated excitation in forward emission. Imhof *et al* [77] showed that paint drying produced easily measurable changes of opto-thermal decay rate in impulse PTR observations in backward emission.

Fig. 14 (a) Schematic diagram of a hemispherical cavity used for lumber moisture content measurement. (b) Moisture related changes in (a) thermal conductivity, (b) thermal effusivity and (c) front surface peak temperature rise in a softwood veneer sample. (From Ref. [93], with permission).

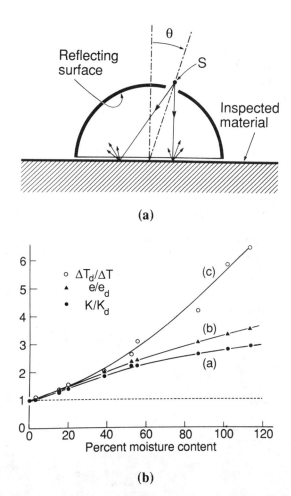

Lumber moisture evaluation was investigated by Cielo *et al* [93]. They used a reflective cavity in contact with the sample, as illustrated in Fig. 14(a), to increase both absorptivity and emissivity of the inspected surface to near unity, thus making the measurements independent of the optical properties of the samples. A step excitation technique was used to measure thermal effusivity, a parameter that was found to be more sensitive to moisture content than thermal diffusivity, as illustrated in Fig. 14(b). Bindra *et al* [94, 95] showed that hydration of human skin can produce measurable effects in PTR impulse response curves. All these investigations used excitation wavelengths that were absorbed by the substrate, but not by the water itself, so that moisture-dependent effects came from changes in infrared absorbance, thermal and geometrical parameters. Higher sensitivity might be achieved by using excitation wavelengths that are strongly absorbed by water, such as the 2.94μm radiation from Er:YAG. Zharov *et al* [96] reported the use of such a laser in impulse PTR experiments, but without investigating moisture content specifically. Our own preliminary results on skin hydration using Spectron Q-switched Er:YAG excitation look promising.

Fig. 15 PTR signals from a float glass excited with 266nm radiation. Decaying signals are seen for the bulk glass (edge) and the float surface, but the non-float surface gives a delayed thermal wave signal, indicating a non-uniform distribution of impurities [99].

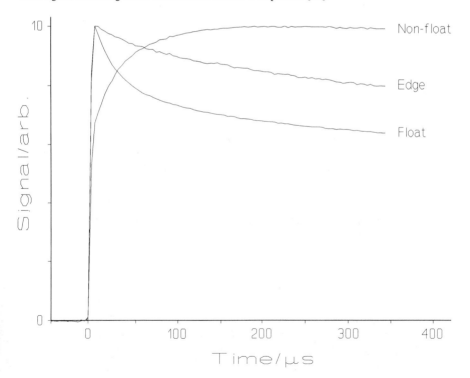

5.3 Other Characterizations

In an elegant experiment, Busse [97] demonstrated that thermal wave transmission could be used to map out changes with depth of structure and composition of case hardened of steel. Potier et al [98] demonstrated the use of PTR in identifying steels of different compositions and surface treatments. Imhof et al [99, 100] observed unusual delayed thermal wave signals in the non-float surface of float glasses, as illustrated in Fig. 15. These indicate a peak in the distribution of impurities, located 10-20µm beneath the surface. Signals from the float surface were found to be stronger and decay more rapidly than those from the bulk glass, observed at a fracture edge, which indicates a high concentration of impurities close to the float surface. Such measurements could be used for characterizing near-surface impurity distributions, or for identification in forensic investigations.

Impulse excitation PTR also offers a unique approach to material characterization through measurements of phase transition dynamics. If the steady sample temperature is close but below a phase transition temperature between two condensed phases, then the energy injected by the excitation pulse may be sufficient for the material near the surface to change phase momentarily. The effects of such reversible phase transitions on the infrared emission are little understood. Imhof et al [101] studied the melting transition of benzophenone, where the effect appeared to stem mainly from the storage and release of latent heat. A study of spectrally resolved infrared transients from a liquid crystal near its nematic to isotropic phase transition revealed a highly complex pattern involving oscillations, including even the transient reduction of thermal emission levels below their equilibrium rates [102]. These were explained tentatively in terms of energy transfer between molecular vibrations induced by changes in lattice structure. In a practical use of phase transition dynamics, Imhof et al [103] showed that the structure of a cosmetic emulsion broke down more rapidly than its timescale of disappearance from the surface of *in-vivo* human skin. Related work showed that the disappearance of substances from *in-vivo* skin could be measured both when the substance itself was the main source of signal [104] and when it modified the skin while itself remaining invisible to the excitation [94, 105].

6. Flaw Detection and Characterization

The research reviewed thus far has been directed towards measuring or characterizing the physical properties of samples. This section deals specifically with the detection and characterization of imperfections. The principle used for this is to detect changes in heat flow that can be related to thermal resistance. Theoretical and numerical studies relating to flaw detection and characterization include Balageas et al [106, 107], Egée et al [108], Aamodt et al [34], and the work of Almond and co-workers [109-114].

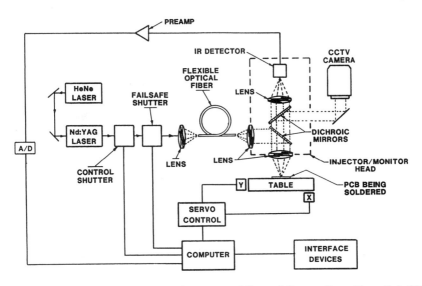

Fig. 16 Automatic solder joint inspection system of Vanzetti Systems Inc. (From Ref. [119], with permission).

Fig. 17 Infrared signatures of solder joints. (From the authors of Ref. [119], (private communication) with permission).

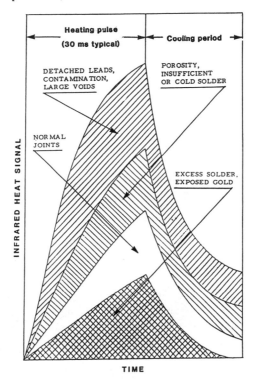

As far back as 1968, Kubiak [115] published details of a PTR NDE system for detecting fatigue cracks and other near-surface defects in metallic aircraft skins. It used a flying spot principle with arc lamp excitation and Ge:Hg infrared detection to map out C-scans on a facsimile recorder, at a rate of typically $5 \text{cm}^2 \text{s}^{-1}$. The optical system used coupled oscillating mirrors to scan the excitation and thermal emission spots in synchronism, with constant velocity and an adjustable offset between them, to give a backward emission measurement geometry with transverse heat flow, analogous to Fig. 1(b). Scanning in the second dimension was achieved by moving the sample. The detection of cracks relied upon the disturbance they produced to the heat flow away from the heated spot. Other work on flaw detection in metals includes the experiments of Busse and co-workers, who used sinusoidal excitation and a variety of measurement methods [75, 116, 117]. Of particular interest was the use of a novel stereoscopic technique with two thermal wave point sources displaced by a known distance, to measure the depth of defects [116]. Maclachlan Spicer *et al* [118] were able to detect disbonds in a sandwich

Fig. 18 PTR impulse response curves for a laminated graphite-epoxy sheet. Curve (a) shows the response of a well bonded region, curves (b) and (c) reveal delaminations. (From Ref. [120], with permission).

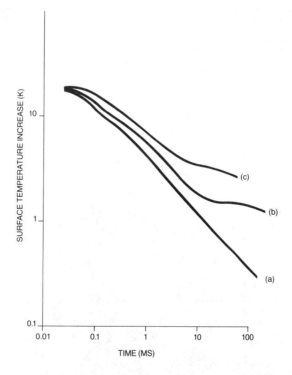

structure of ceramic on aluminium, using step excitation with a scanning line source to map variations across the sample.

Vanzetti and Traub [119] reported an automatic solder joint inspection system using a hybrid step/impulse PTR technique in backward emission. The system illustrated in Fig. 16 used focused Nd:YAG laser pulses of 30ms duration and 1.06μm wavelength for excitation. It measured heating/cooling curves, such as those illustrated in Fig. 17, with a liquid nitrogen cooled InSb detector. The signal processing was automatic and classified the solder joints according to known characteristics of common faults, such as insufficient solder and internal voids, at rates up to 10 joints per second.

Cielo [120] developed an impulse technique in backward emission for detecting disbonds in stratified materials, including thermally sprayed metallurgical coatings and laminates. These were shown to affect the $t^{-1/2}$ decay law for thick, homogeneous materials in ways that depend on changes in effusivity and thermal resistance. Experimental curves for a graphite-epoxy laminated sheet consisting of four layers, each of 125μm thickness, are shown in Fig. 18. The lower curve is characteristic of a well bonded area, whereas the upper curves show regions of de-lamination, whose depth can be deduced from their ≈10ms thermal diffusion time. Thinner films require higher time resolution for testing, as shown in the work of Tam *et al* [81] on printer ribbons, illustrated in Fig. 19. The conventional impulse PTR apparatus was adapted to measure transients from film moving between two spools, thus making it suitable for rapid testing. Typical transients for good and faulty ribbon are shown in Fig. 19(b). The use of opto-thermal decay rate in such non-destructive tests renders the measurement insensitive to instrumental effects. The disadvantage is that such decay rates are hybrid parameters, depending on both optical and thermal properties of the sample. Bonding between layers can also be studied with sinusoidal excitation, as shown by a number of authors [75, 87, 89, 108, 121, 122]. An interesting variant of sinusoidal modulation, namely random modulation, was used by Egée *et al* to study adhesion defects between sprayed ceramic coatings and metal substrates. The technique attempts to combine the advantages of sinusoidal excitation for depth selectivity and low thermal stresses with the speed and range of depth probing of impulse excitation. Fig. 20 shows the experimental curves obtained with a sample consisting of an aluminium substrate and a 200μm zirconia coating.

The above techniques are generally insensitive to cracks normal to the sample surface, unless rapid flying spot scanning of a finely focused beam is used. Another approach was explored by Enguehard *et al* [59], who showed that the convergent thermal wave technique can detect such faults.

7. PTR NDE Measurement Systems

There are as yet few examples where PTR has made the transition

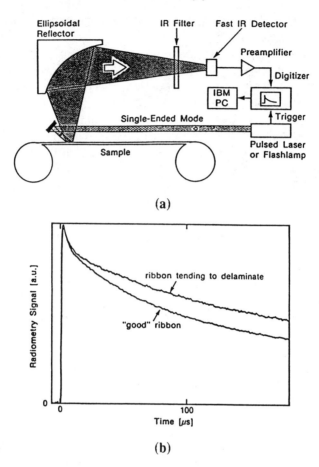

Fig. 19 (a) Impulse PTR apparatus for rapidly testing thin films moving between spools. (b) Effect of delamination is shown as a decrease of signal decay rate. (From Ref. [81], with permission).

Fig. 20 Measured phase angle response of a zirconia coated specimen with and without adhesion defect, measured using randomly modulated excitation. (From Ref. [123], with permission).

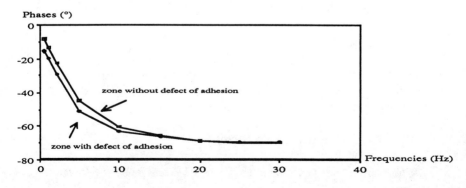

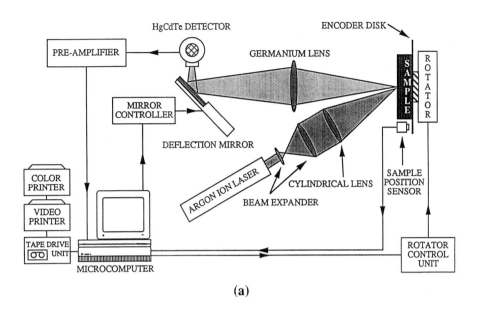

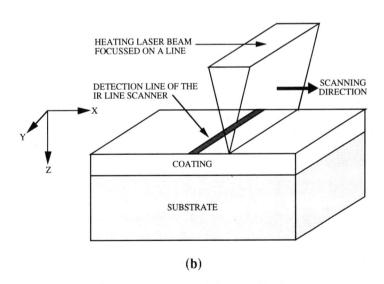

Fig. 21 (a) Schematic diagram of a line scanning PTR measurement system; (From Ref. [124], with permission). (b) Details of excitation and detection geometry. (From Ref. [125], with permission).

from laboratory to industrial production environment. To do so, PTR NDE systems must find design compromises between the often conflicting requirements of the fundamental physics of the measurement process on the one hand and rugged, no fuss functionality on the other. Of course, some NDE tasks can be performed by taking samples to a laboratory, to be tested by an expert. But the ultimate application aim of the research reviewed in this chapter must be to come out of the laboratory and become useful in the hands of non-experts.

In much industrial testing, the ability to measure an image in some form or other is paramount. One method of doing this is video thermography, which has made rapid progress towards routine use, largely because of advances in camera design. However, video thermography is outside the scope of this chapter. Of course, many PTR experiments reviewed in this chapter are capable of producing an image by mapping out the measurements over a grid of points, but most are too slow for practical use.

The only realistic alternative to the video camera for PTR imaging is flying spot excitation, as explored by Kubiak (See Sect. III(6)) and the Helsinki group. A line scanning system for mapping faults in plasma sprayed coatings [124, 125] is illustrated in Fig. 21. The laser beam was expanded by means of an anamorphic lens system to form a line on the sample surface. During the measurement, the image point of the detector was scanned by means of a deflection mirror over the sample surface in the y-direction, while the sample was moved simultaneously in the x-direction. The signal was digitized and processed by a microcomputer, which also controlled the experimental parameters, including time delay between heating and detection. Fig. 22 shows a scanned image of a steel sample coated with a 200µm thick chromium oxide layer and containing an artificial adhesion defect. The image of 8×6mm^2 was signal averaged eight times during a total measurement time of 42s. The authors pointed out that the defect map is thermal in nature, since the sample was found to be optically uniform in both visible and infrared illumination. The authors also described a system using point rather than line focussing [126, 127], with the added feature of allowing the optical and thermal images to be recorded simultaneously.

Both the above systems are still laboratory bound. Varis *et al* [128] recently described a purpose designed instrument for industrial maintenance and testing, which is shown schematically in Fig. 23. It consisted of a hand-held measurement head connected to a remote instrument rack. The head contained a liquid nitrogen cooled detector, optics and scanning mirrors. The rack contained a cw-Nd:YAG laser, signal processing electronics and control computer. An optical fiber delivered the excitation to the head, where it was formed into a line source. Images were collected in a similar manner to the system in Fig. 21. Its measurement capabilities are illustrated in Fig. 24. The test sample consisted of an optically homogeneous composite plate with several layers of carbon and glass fibers, as illustrated schematically in

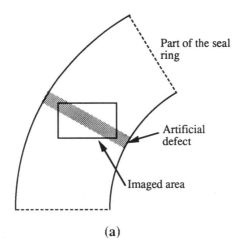

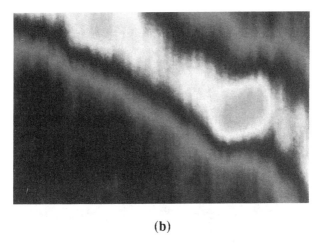

Fig. 22 Scanned image of an adhesion defect in a plasma sprayed coating, measured using the line scanning apparatus of Fig. 21. (From Ref. [124], with permission).

Fig. 24(b). The scanned image, Fig. 24(a) shows an inner resolved structure caused by the higher thermal conductivity of the carbon fibres. The dimensions of the image were $60 \times 20 \text{mm}^2$ and the measurement time was a remarkable 6s.

8. Commercial PTR NDE Instrumentation

If conventional flash diffusivity instrumentation is excluded as being too far removed from practical NDE, there remain three commercial PTR NDE measurement systems on the market at present. The solder joint inspection system [119] of Vanzetti Systems Inc. has already been described

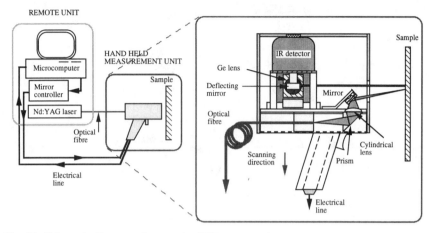

Fig. 23 Schematic diagram of a scanning PTR system with hand-held measurement head, for industrial testing applications. (From Ref. [128], with permission).

Fig. 24 (a) Scanned image of a carbon and glass fiber composite panel, measured with the hand-held PTR scanner of Fig. 23. (b) Schematic diagram of the internal structure of the composite tested above. (From Ref. [128], with permission).

in Section III(6). The P.S. system of Phototherm Dr. Petry GmbH comprises a relatively compact measuring head connected to a signal processing electronics and display unit. In its standard form, it uses a sinusoidally modulated diode laser of 830nm wavelength, focused to a spot of approximately 1mm diameter, with Joule-Thomson cooled InSb infrared detection, and offers automatic, non-invasive film thickness measurement for a wide range of material combinations, with measurement times in the range 1-5s. The *ALADIN* Thermo Microscope of Siemens AG has raster scanning of a laser beam built into the measuring head. The brochure shows many application of examples, but gives no details of how the measurements are performed or evaluated.

IV. CONCLUSIONS

This review shows that progress in PTR is accelerating towards practical applications. There continues to be a need for fundamental theoretical and experimental work, but some areas are clearly ripe for exploitation. It must not be forgotten that fundamental PTR research can sometimes progress towards applications in ways that move away from PTR *per se*, with non-optical excitation methods or video thermographic imaging for example. Therefore, it is important not to regard the work reviewed in this chapter in isolation from other optical and thermal techniques.

V. REFERENCES

1. A.C. Tam, Rev. Mod. Phys. **58**, 381 (1986).

2. *Principles & Perspectives of Photothermal & Photoacoustic Measurements*, (A. Mandelis, Ed.) Progress in Photothermal & Photoacoustic Science & Technology, **1** (Elsevier, New York, 1992).

3. A.C. Boccara and D. Fournier, *Principles & Perspectives of Photothermal & Photoacoustic Measurements*, (A. Mandelis, Ed.) Progress in Photothermal & Photoacoustic Science & Technology, **1**, 3 (Elsevier, New York, 1992).

4. G. Busse and H.G. Walther, *Principles & Perspectives of Photothermal & Photoacoustic Measurements*, (A. Mandelis, Ed.) Progress in Photothermal & Photoacoustic Science & Technology, **1**, 205, (Elsevier, New York, 1992).

5. M. Mundiasa and A. Mandelis, *Principles & Perspectives of Photothermal & Photoacoustic Measurements*, (A. Mandelis, Ed.) Progress in Photothermal & Photoacoustic Science & Technology, **1**,

299 (Elsevier, New York, 1992).

6. D. Balageas, Tech. de l'Ing. **10**, R2 955-1 (1986).

7. D.L. Balageas, High Temp. High Press. **21**, 85 (1989).

8. G. Busse, Journal de Physiqe, **44**(C6) 427 (1983).

9. A.C. Tam, *Photoacoustic and Photothermal Phenomena in Semiconductors*, (A. Mandelis, Ed.) 188 (Elsevier North Holland, New York, 1987).

10. G. Busse and D. Vergne, *Deutsche Forschungsgesellschaft fuer Oberflaechenbehandlung e.V.* 362 (1989).

11. G. Busse, *Physical Acoustics*, (O. Leroy and M.A. Breazeale, Eds.) 31 (Plenum, New York, 1991).

12. D. Balageas, Revue Générale de Thermique, No 356-7, 483 (1991).

13. R.E. Imhof, C.J. Whitters & D.J.S. Birch, *Photoacoustic & Photothermal Phenomena II*, (J.C. Murphy, J.W. Maclachlan-Spicer, L. Aamodt & B.S.H. Royce, Eds.) Springer Series in Optical Sciences, **62**, 46-54 (Springer, Berlin, 1990).

14. P.M. Patel, D.P. Almond and J.D. Morris, Europ. J. NDT, **1**, 64 (1991).

15. *Proceedings of the 4th International Topical Meeting of Photoacoustics, Thermal, and Related Sciences*, Can. J. Phys. **64** (1986).

16. *Photoacoustic & Photothermal Phenomena*, (P.Hess & J.Petzl, Eds.) Springer Series in Optical Sciences, **58** (Springer-Verlag, Berlin, 1988).

17. *Photoacoustic & Photothermal Phenomena II*, (J.C. Murphy, J.W. Maclachlan-Spicer, L. Aamodt & B.S.H. Royce, Eds.) Springer Series in Optical Sciences, **62** (Springer-Verlag, Berlin, 1990).

18. *Photoacoustic & Photothermal Phenomena III*, (D. Bićanić, Ed.) Springer Series in Optical Sciences, **69** (Springer-Verlag, Berlin, 1992).

19. H.S. Carslaw and J.C. Jaeger *Conduction of Heat in Solids* (Oxford University Press, 1959).

20. H.P. Baltes, *Progress in Optics*, (E. Wolf Ed.), **13** (New York 1976).

21. M.G. Dreyfus, Appl. Opt. **2**, 1113 (1963).

22. P.K. Kuo, L.J. Inglehart, L.D. Favro and R.L. Thomas, *Proc. IEEE Ultrasonics Symposium*, 788 (1981).

23. P.K. Kuo and L.D. Favro, Appl. Phys. Lett. **40**, 1012 (1982).

24. R.E. Imhof, D.J.S. Birch, F.R. Thornley, J.R. Gilchrist & T.A. Strivens, J.Phys.E: Sci.Instrum. **17**, 521 (1984).

25. P-E. Nordal and S.O. Kanstad, Physica Scripta, **20**, 659 (1979).

26. C.A. Bennett Jr. and R. R. Patty, Appl. Opt. **21**, 49 (1982).

27. M. Lax, J. Appl. Phys. **48**, 3919 (1977).

28. M.F. Modest and H. Abakians, J. Heat Trans. **108**, 597 (1986).

29. H.J. Vidberg, J. Jaarinen and D.O. Riska, Can. J. Phys. **64**, 1178 (1986).

30. A. Mandelis, S.B. Peralta and J. Thoen, J. Appl. Phys. **70**, 1761 (1991).

31. A. Mandelis, E. Schoubs, S.B. Peralta and J. Thoen, J. Appl. Phys. **70**, 1771 (1991).

32. M. Munidasa, T.C. Ma, A. Mandelis, S.K. Brown and L. Mannik, J. Mat. Sci. Eng. **A159**, 111 (1992).

33. T.C. Ma, M. Munidasa and A. Mandelis, J. Appl. Phys. **71**, 6029 (1992).

34. L.C. Aamodt, J.W. Maclachlan and J.C. Murphy, J. Appl. Phys. **68**, 6087 (1990).

35. R.D. Tom, E.P. O'Hara and D. Benin, J. Appl. Phys. **53**, 5392 (1982).

36. L.C. Aamodt and J.C. Murphy, Can. J. Phys. **64**, 1221 (1986).

37. R. Tilgner, J. Baumann and M. Beyfuss, Can. J. Phys. **64**, 1287 (1986).

38. J. Baumann and R. Tilgner, Can. J. Phys. **64**, 1291 (1986).

39. M. Beyfuss, R. Tilgner and J. Baumann, *Photoacoustic & Photothermal Phenomena*, (P.Hess & J.Petzl, Eds.) Springer Series in Optical Sciences, **58**, 392 (Springer-Verlag, Berlin, 1988).

40. R.L. Rudkin, R.J. Jenkins and W.J. Parker, Rev. Sci. Instrum. **33**, 21 (1962).

41. H.W. Deem and W.D. Wood, Rev. Sci. Instrum. **33**, 1107 (1962).

42. R. Taylor, J. Sci. Instrum. **13**, 1193 (1980).

43. I.G. Korshunov, A.N. Mezentsev, A.D. Ivliev and V.N. Gorbatov, Tep. Vys. Temp. **27**, 63 (1989).

44. C.L. Choy, W.P. Leung and Y.K. Ng, J. Polymer Sci. B: Polymer Phys. **25**, 1779 (1987).

45. N. Tsutsumi and T. Kiyotsukuri, Appl. Phys. Lett. **52**, 442 (1988).

46. N. Tsutsumi, T. Takizawa and T. Kiyotsukuri, Polymer **31**, 1925 (1990).

47. H.P.R. Frederikse, A. Feldman and X.T. Ying, *Photoacoustic & Photothermal Phenomena II*, (J.C. Murphy, J.W. Maclachlan-Spicer, L. Aamodt & B.S.H. Royce, Eds.) Springer Series in Optical Sciences, **62**, 130 (Springer-Verlag, Berlin, 1990).

48. A.W. Schultz, Appl. Opt. **7**, 1845 (1968).

49. P. Cielo, S. Dallaire, G. Lamonde and S. Johar, Can. J. Phys. **64**, 1217 (1986).

50. W.P. Leung and A.C. Tam, Opt. Lett. **9**, 93 (1984).

51. W.P. Leung and A.C. Tam, J. Appl. Phys. **56**, 153 (1984) and **58**, 1087 (1985).

52. W.P. Leung and A.C. Tam, J. Appl. Phys. **64**, 2227 (1988).

53. D.P. Cobranchi, N.F. Leite, J. Isak, S.J. Komorowski, A. Gerhard and E.M. Eyring, *Photoacoustic & Photothermal Phenomena II*, (J.C. Murphy, J.W. Maclachlan-Spicer, L. Aamodt & B.S.H. Royce, Eds.) Springer Series in Optical Sciences, **62**, 328 (Springer-Verlag, Berlin, 1990).

54. H.P.R. Frederikse and X.T. Ying, Appl. Opt. **27**, 4672 (1988).

55. M. Beyfuss and J. Baumann, *Non-destructive Characterization of Materials*, (P. Hoeller, V. Hank, G. Dobmann, C.O. Ruud and R.E. Green, Eds.) 807 (Springer, Berlin, 1989).

56. M. Beyfuss, H. Reichl and J. Baumann, *Photoacoustic & Photothermal Phenomena III*, (D. Bićanić, Ed.) Springer Series in Optical Sciences, **69**, 692 (Springer-Verlag, Berlin, 1992).

57. P. Cielo, L.A. Utracki and M. Lamontagne, Can. J. Phys. **64**, 1172 (1986).

58. F. Enguehard, D.M. Boscher, A.A. Dénom and D.L. Balageas, Mat. Sci. & Eng. **B5**, 127 (1990).

59. F. Enguehard, A.A. Dénom and D.L. Balageas, *Photoacoustic & Photothermal Phenomena III*, (D. Bićanić, Ed.) Springer Series in Optical Sciences, **69**, 537 (Springer-Verlag, Berlin, 1992).

60. P. Cielo, F. Nadeau and M. Lamontagne, Ultrasonics **23**, 55 (1985).

61. S.O. Kanstad and P-E. Nordal, Powder. Technol. **22**, 133 (1978).

62. S.O. Kanstad and P-E. Nordal, Appl. Surf. Sci. **6**, 372 (1980).

63. S.O. Kanstad and P-E. Nordal, Phys. Technol. **11**, 142 (1980).

64. P-E. Nordal and S.O. Kanstad, Appl. Phys. Lett. **38**, 486 (1981).

65. S.O. Kanstad and P-E. Nordal, Can. J. Phys. **64**, 1155 (1986).

66. A.C. Tam, Infrared Phys. **25**, 305 (1985).

67. R.E. Imhof, C.J. Whitters, D.J.S. Birch & F.R. Thornley, J.Phys.E: Sci.Instrum. **21**, 115-7 (1988).

68. R.E. Imhof, C.J. Whitters, D.J.S. Birch & F.R. Thornley,

Photoacoustic & Photothermal Phenomena, (P.Hess & J.Petzl, Eds.) Springer Series in Optical Sciences, **58**, 503 (Springer-Verlag, Berlin, 1988).

69. R.W. Jones and J.F. McClelland, Anal. Chem. **61**, 650 (1989).

70. R.W. Jones and J.F. McClelland, Anal. Chem. **61**, 1810 (1989).

71. R.R. Anderson, H. Beck, U. Bruggemann, W. Farinelli, S.L. Jacques and J.A. Parrish, Appl. Opt. **28**, 2256 (1989).

72. R.E. Imhof, D.J.S. Birch, M.M. Moksin, J.F. Webb, P.H. Willson & T.A. Strivens, British J.NDT, **33**, 172 (1991).

73. G. Busse, Infrared Phys. **20**, 419 (1980).

74. S.J. Sheard and M.G. Somekh, Appl. Phys. Lett. **53**, 2715 (1988).

75. G. Busse, D.Vergne and B. Wetzel, *Photoacoustic & Photothermal Phenomena*, (P.Hess & J.Petzl, Eds.) Springer Series in Optical Sciences, **58**, 427 (Springer-Verlag, Berlin, 1988).

76. W. Karpen, A. Bohnacker, H.G. Walther, K. Friedrich, U. Seidel and G. Busse, *Photoacoustic & Photothermal Phenomena III*, (D. Bicanic, Ed.) Springer Series in Optical Sciences, **69**, 248 (Springer-Verlag, Berlin, 1992).

77. R.E. Imhof, C.J. Whitters & D.J.S. Birch, Materials Science & Engineering, **B5**, 113-7 (1990).

78. R.E. Imhof, F.R. Thornley, J.R. Gilchrist & D.J.S. Birch, J.Phys.D: Appl.Phys. **19**, 1829 (1986).

79. P. Cielo and S. Dallaire, J. Mater. Eng. **9**, 71 (1987).

80. M.E. Pita de Jesus, R.E. Imhof, D.J.S. Birch, J.F. Webb, P.H. Willson & T.A. Strivens, *Photoacoustic & Photothermal Phenomena III*, (D. Bicanic, Ed.) Springer Series in Optical Sciences, **69**, 213 (Springer-Verlag, Berlin, 1992).

81. A.C. Tam, W.P. Leung and H. Sonntag, *Photoacoustic & Photothermal Phenomena III*, (D. Bicanic, Ed.) Springer Series in Optical Sciences, **69**, 672 (Springer-Verlag, Berlin, 1992).

82. W.P. Leung and A.C. Tam, J. Appl. Phys. **63**, 4505 (1988).

83. A.C. Tam and H. Sonntag, Appl. Phys. Lett. **49**, 1761 (1986).

84. A.C. Tam, H. Sonntag and W.P. Leung, *Photoacoustic & Photothermal Phenomena*, (P.Hess & J.Petzl, Eds.) Springer Series in Optical Sciences, **58**, 419 (Springer-Verlag, Berlin, 1988).

85. M. Beyfuss, J. Baumann and R. Tilgner, *Photoacoustic & Photothermal Phenomena II*, (J.C. Murphy, J.W. Maclachlan-Spicer, L. Aamodt & B.S.H. Royce, Eds.) Springer Series in Optical Sciences, **62**, 17 (Springer-Verlag, Berlin, 1990).

86. J.D. Morris, D.P. Almond, P.M. Patel and H. Reiter, *Photoacoustic & Photothermal Phenomena*, (P.Hess & J.Petzl, Eds.) Springer Series in Optical Sciences, **58**, 424 (Springer-Verlag, Berlin, 1988).

87. P.M. Patel and D.P. Almond, J. Mat. Sci. **20**, 955 (1985).

88. M. Heuret, C. Bissieux, L. Pincon, P. Egée, G. Kurka and J. Danroc, Surf. & Coat. Techn. **45**, 325 (1991).

89. B. Rief, G. Busse and P. Eyrer, *Photoacoustic & Photothermal Phenomena*, (P.Hess & J.Petzl, Eds.) Springer Series in Optical Sciences, **58**, 447 (Springer-Verlag, Berlin, 1988).

90. R.E. Imhof, D.J.S. Birch, F.R. Thornley, J.R. Gilchrist & T.A. Strivens, J.Phys.D: Appl.Phys. **18**, L103 (1985).

91. J.C. Krapez, P. Cielo, K. Cole and G. Vandreuil, J. Therm. Anal. **32**, 1859 (1987).

92. W.P. Leung and A.C. Tam, Appl. Phys. Lett. **51**, 2085 (1987).

93. P. Cielo, J.C. Krapez and M. Lamontagne, Rev. Phys. Appl. **23**, 1565 (1988).

94. R.M.S. Bindra, G.M. Eccleston, R.E. Imhof & D.J.S. Birch, *Photoacoustic & Photothermal Phenomena III*, (D. Bicanic, Ed.) Springer Series in Optical Sciences, **69**, 95 (Springer, Berlin, 1992).

95. R.M.S. Bindra, G.M. Eccleston, R.E. Imhof and D.J.S. Birch, Pred. Perc. Pen: Meths. Meas. & Mod. **2**, 628 (1992).

96. V.P. Zharov, A.A. Kozlov, Yu.O. Litvinov, B.V. Zubov and A.A. Savelyev, *Photoacoustic & Photothermal Phenomena II*, (J.C. Murphy, J.W. Maclachlan-Spicer, L. Aamodt & B.S.H. Royce, Eds.) Springer Series in Optical Sciences, **62**, 463 (Springer-Verlag, Berlin, 1990).

97. G. Busse, *Photoacoustic, Photothermal and Photochemical Processes at Surfaces and in Thin Films*, (P.Hess, Ed.) Springer Ser. Topics in Current Physics, **47**, 251 (Springer-Verlag, Berlin, 1989).

98. F. Potier, E. Van Schel, C. Bissieux, J.L. Beaudoin, M. Heuret, M. Régalia and M. Egée, *Actes du Colloque de Thermique Progrés et Défis Actuels* (ISITEM Nantes, 1990).

99. R.E. Imhof, D.J.S. Birch, R.M.S. Bindra, P.H. Willson, J. Locke & D.G. Sanger, J.Phys.D: Applied Physics, **24**, 2067 (1991).

100. R.E. Imhof, R.M.S. Bindra, D.J.S. Birch, P.H. Willson, J. Locke, D.J. Sanger & N.D. Cowan, *Photoacoustic & Photothermal Phenomena III*, (D. Bićanić, Ed.) Springer Series in Optical Sciences, **69**, 209 (Springer-Verlag, Berlin, 1992).

101. R.E. Imhof, F.R. Thornley, J.R. Gilchrist & D.J.S. Birch, Appl.Phys.B, **43**, 23 (1987).

102. R.E. Imhof, C.J. Whitters & D.J.S. Birch, Infrared Phys. **29**, No.2-4, 433-40 (1989).

103. R.E. Imhof, C.J. Whitters & D.J.S. Birch, Clinical Materials, **5**, 2-4, 271 (1990).

104. R.E. Imhof, C.J. Whitters & D.J.S. Birch, Physics in Medicine & Biology, **35**, 95 (1990).

105. R.M.S. Bindra, R.E. Imhof, G.M. Eccleston and D.J.S. Birch, *SPIE Proc.* **1643**, 299 (1992).

106. D.L. Balageas, J.C. Krapez and P. Cielo, J. Appl. Phys. **59**, 348 (1986).

107. D.L. Balageas, A.A. Dénom and D.M. Boscher, Mater. Eval. **45**, 461 (1987).

108. M. Egée, R. Dartois, J. Marx and C. Bissieux, Can. J. Phys. **64**, 1297 (1986).

109. P.M. Patel, D.P. Almond and H. Reiter, *Photoacoustic & Photothermal Phenomena*, (P.Hess & J.Petzl, Eds.) Springer Series in Optical Sciences, **58**, 430 (Springer-Verlag, Berlin, 1988).

110. J.D. Morris, D.P. Almond, P.M. Patel and H. Reiter, *Photoacoustic & Photothermal Phenomena II*, (J.C. Murphy, J.W. Maclachlan-Spicer, L. Aamodt & B.S.H. Royce, Eds.) Springer Series in Optical Sciences, **62**, 71 (Springer-Verlag, Berlin, 1990).

111. S.K. Lau, D.P. Almond and P.M. Patel, *Photoacoustic & Photothermal Phenomena II*, (J.C. Murphy, J.W. Maclachlan-Spicer, L. Aamodt & B.S.H. Royce, Eds.) Springer Series in Optical Sciences, **62**, 78 (Springer-Verlag, Berlin, 1990).

112. P.M. Patel, S.K. Lau and D.P. Almond, *Photoacoustic & Photothermal Phenomena III*, (D. Bićanić, Ed.) Springer Series in Optical Sciences, **69**, 235 (Springer-Verlag, Berlin, 1992).

113. P.M. Patel, S.K. Lau and D.P. Almond, *Photoacoustic & Photothermal Phenomena III*, (D. Bićanić, Ed.) Springer Series in Optical Sciences, **69**, 238 (Springer-Verlag, Berlin, 1992).

114. P.M. Patel, S.K. Lau and D.P. Almond, *Photoacoustic & Photothermal Phenomena III*, (D. Bićanić, Ed.) Springer Series in Optical Sciences, **69**, 241 (Springer-Verlag, Berlin, 1992).

115. E.J. Kubiak, Appl. Opt. **7**, 1743 (1968).

116. G. Busse and K.F. Renk, Appl. Phys. Lett. **42**, 366 (1983).

117. G. Busse, B. Rief and P. Eyerer, Can. J. Phys. **64**, 1195 (1986).

118. J.W. Maclachlan Spicer, W.D. Kerns, L.C. Aamodt and J.C. Murphy, *Photoacoustic & Photothermal Phenomena II*, (J.C. Murphy, J.W. Maclachlan-Spicer, L. Aamodt & B.S.H. Royce, Eds.) Springer Series in Optical Sciences, **62**, 55 (Springer-Verlag, Berlin, 1990).

119. R. Vanzetti and A.C. Traub, Proc. 7th Annual Seminar, *Soldering Technology and Product Assurance*, Code 3681, 1 (Naval Weapons Center, China Lake, California, 1983).

120. P. Cielo, J. Appl. Phys. **56**, 230 (1984).

121. J. Corbett, M.B.C. Quigley, B. Hart and B.L. Smith, *Photoacoustic &*

Photothermal Phenomena, (P.Hess & J.Petzl, Eds.) Springer Series in Optical Sciences, **58**, 440 (Springer-Verlag, Berlin, 1988).

122. M. Heuret, E. Van Schel and M. Egée, Mat.Sci.& Eng **B5**, 119 (1990).

123. P. Egée, E. Mérienne, K. Hakem, M. Heuret and M. Egée, *Photoacoustic & Photothermal Phenomena III*, (D. Bićanić, Ed.) Springer Series in Optical Sciences, **69**, 205 (Springer-Verlag, Berlin, 1992).

124. J. Hartikainen, Rev. Sci. Instrum. **60**, 1334 (1989).

125. R. Lehtiniemi, J. Hartikainen, J. Rantala, J. Varis and M. Luukkala, Rev. Progr. in Quant. NDE, preprint (1991).

126. J. Hartikainen, Appl. Phys. Lett. **55**, 1188 (1989).

127. J. Hartikainen and M. Luukkala, *Photoacoustic & Photothermal Phenomena II*, (J.C. Murphy, J.W. Maclachlan-Spicer, L. Aamodt & B.S.H. Royce, Eds.) Springer Series in Optical Sciences, **62**, 496 (Springer-Verlag, Berlin, 1990).

128. J. Varis, J. Hartikainen, R. Lehtiniemi, J. Rantala, and M. Luukkala, *Photoacoustic & Photothermal Phenomena III*, (D. Bicanic, Ed.) Springer Series in Optical Sciences, **69**, 565 (Springer-Verlag, Berlin, 1992).

SPATIALLY RESOLVED DETECTION OF MICROWAVE ABSORPTION IN FERRIMAGNETIC MATERIALS

J. Pelzl and O. von Geisau

Institut für Experimentalphysik, AG Festkörperspektroskopie, Ruhr-Universität Bochum, D-44780 Bochum, FRG

I. INTRODUCTION 238

II. EXPERIMENTAL ASPECTS 240
   1. Microwave Resonance Absorption 240
   2. Detection Schemes 245
   3. Experimental Arrangements 247
   4. Sensitivity and Spatial Resolution 250

III. APPLICATIONS TO FERRIMAGNETIC MATERIALS 261
   1. Magnetic Properties of Ferrimagnetic Materials 262
   2. Magnetic Excitations 263
   3. Microwave Density Distribution in Soft Ferrite Slabs ... 266
   4. Magnetostatic Modes in Yttrium Iron Garnet (YIG) 288

IV. SUMMARY 311

I. INTRODUCTION

Photothermal methods are based on the detection of the heat generated in the material in the course of absorption of radiation. The outstanding significance of these methods is due to the conversion process of the radiative energy to heat in the matter and on the succeeding time evolution of the spatial distribution of the temperature field in the sample. Investigations devoted to the first process provide information on the non-radiative decay channels following an absorption process which can be separated by a photothermal experiment from other mechanisms of energy dissipation [1-5]. The second attribute can be used to investigate the spatial distribution of the heat sources or the spatial inhomogeneities of properties which are governing the thermal transport. This application has become one of the most important potentials of photothermal methods in the field of non-destructive evaluation and characterization of materials [2,6,7].

A significant feature of photothermal methods is the time dependence of the heat generation process which can be pulse shaped or oscillatory in time giving rise to heat pulses or to thermal waves travelling through the specimen. The time delay of the thermal wave reaching e.g. the surface of the specimen is a direct measure of the distance of the absorbing center inside the material. Besides this spatial discrimination ability of the photothermal signal, the amplitude of the thermal wave yields information on the absorbing strength which is governed by the interaction of the exciting radiation with matter. The nature of the radiation-induced fundamental excitations depends on the kind and on the energy of the radiation impinging on the sample.

The types of radiation used nowadays for the excitation of thermal waves cover a wide spectrum of energies ranging from μeV to MeV and they comprise particle radiation as well as electromagnetic radiation (Table 1.1). The charged particles like the electrons and the ions transfer their kinetic energy by collision processes to the lattice. As in the case of x-rays this process can also involve core electron excitations. The efficiency of the conversion of radiation energy to heat decreases at higher energies of the particles or electromagnetic waves as an increasing portion of the energy is disposed of for the non-reversible displacements of lattice ions. Visible and infrared light from lasers are most frequently used for the excitation of thermal waves. The first pioneering applications of the photothermal technique comprised the depth profiling of the optical absorptivity in various materials [8] and the separation of non-radiative decay processes [9]. The electronic levels populated by the absorption of photons are depopulated either radiatively or non-radiatively and the efficiency for the heat generation depends on the ratio of these two processes. At again lower energies in the far-infrared and microwave wavelength region (0.1 mm $\leq \lambda \leq 10$ cm) vibrational levels can be excited by photon absorption. However, the outstanding significance of the microwaves in spectroscopic research is substantiated by their direct response to magnetic excitations of spin and orbital magnetic moments. Radiowaves due to their much lower energy couple to nuclear magnetic moments.

With regard to a spatially resolved spectroscopy the microwaves and the radiowaves suffer from the fact that these radiations cannot be focused to a small spot like light and particle radiation. Therefore, local resolution has to be achieved in the detection process. The commonly used technique for spatially resolved radiowave absorption, the nuclear magnetic resonance tomography, applies a magnetic field gradient to impose a spatially varying resonance condition across the specimen [10]. This technique has also been extended to the microwave region [11,12] but the spatial resolution is restricted to the mm-range. Other spectroscopic techniques that are able to measure local absorption effects are based on the interaction of a light beam with the sample. By means of Brillouin light scattering [13] and high-frequency Kerr effect [14] or Faraday effect [15] coherent magnons excited by microwaves could be measured in optically transparent samples such as in yttrium iron garnet. The main drawbacks of these methods are that they only provide a lateral resolution and that they require high quality surfaces in order to achieve a reasonable sensitivity. Therefore, up to now, these optical methods have only been used in a few cases to obtain locally resolved information [16]. On the other hand, a "time of flight" measurement of thermal waves offers the unique tool to obtain a depth as well as a lateral dependent image of the microwave absorptivity with a spatial resolution which is two orders of magnitude higher than that of the gradient techniques. In addition, the photothermal methods are much easier to handle

Table 1 Radiation used for the excitation of photothermal signals.

Particle Radiation		Electromagnetic Radiation					
ions	electrons	x-rays	visible	infrared		microwaves	radiowaves
FOCUSING ABILITIES						NO FOCUSING ABILITIES	
	core electron excitation		valence electron excitation				
collision process				vibrational excitation			
						MAGNETIC EXCITATION	

experimentally than the pure optical detection schemes. The photothermal signal level, in general, is lower than in the case of a conventional microwave detection which is a consquence of the confinement to a small part of the sample necessary to achieve the spatial discrimination. The ability of spatial resolution makes the distinction between photothermal and bolometric techniques. The latter have a long history in magnetic resonance absorption. The first resonance absorption of electromagnetic energy via thermal response was proposed already in 1936 by Gorter [17]. Nowadays the bolometric detection of the paramagnetic resonance is frequently used at low temperature as the most sensitive method [18]. Photothermal methods, on the other hand, are commonly performed at room temperature and in the time or frequency domain as the depth resolution is based on the time dependence of the heat transport.

The most favorable systems for a photothermal inspection of microwave absorption are materials with a high concentration of microwave absorbing centers such as undiluted paramagnets or ferro- or ferrimagnetic materials and thin films and layers where the signal is generated in the near surface region. Consequently, the first microwave resonance measurements using photothermal detection have been performed on ferromagnetic iron layers [19,20] and on DPPH powder which is the spectroscopic standard for electron paramagnetic resonance [21]. Using photoacoustic detection [22] and the mirage effect [23] the ability of the photothermal methods for a spatially resolved investigation of magnetic properties has been demonstrated. The recent introduction of a novel technique, the photothermally modulated microwave resonance absorption, has allowed the sensitivity and the local resolution to be improved considerably [24].

In this chapter we first briefly review the basic concepts of the ferromagnetic resonance (FMR) and summarize the experimental arrangements and conditions of the various photothermal detection schemes in comparison to the conventional recording. The second part of the chapter deals with some recent applications of the photothermally detected and photothermally modulated FMR to the exploration of the spatial variation of magnetic excitations in ferrite slabs and spheres. For results obtained from other materials than ferrites we would like to refer the reader to earlier reports [25,26].

II. EXPERIMENTAL ASPECTS

1. Microwave Resonance Absorption

In a magnetic material which is exposed to microwaves the high-frequency magnetic field of the microwave radiation induces a time-dependent magnetization m. In the limit of linear response m is proportional to the magnetic field b

$$m = \tilde{\chi} b. \tag{1}$$

$\tilde{\chi}$ is the so-called magnetic susceptibility tensor. For a time dependent oscillatory driving field b the magnetization m may be shifted in phase with respect to the oscillating field. Therefore, the high-frequency susceptibility in general is a complex quantity:

$$\tilde{\chi}(\omega) = \tilde{\chi}'(\omega) - i\tilde{\chi}''(\omega) \tag{2}$$

Sometimes it is more convenient to consider the permeability tensor $\tilde{\mu}$ which is related to the magnetic susceptibility tensor by:

$$\tilde{\mu} = \tilde{1} + \tilde{\chi} \tag{3}$$

The real part of the susceptibility or permeability which is in-phase with the high-frequency magnetic field describes the dispersion caused by the material. The out-of-phase imaginary components respresent the dissipative effects.
As the response of a magnetic moment to a time-dependent magnetic field depends on the relative orientation of the two vectors, the magnetic moments have to be aligned in an external static magnetic field B_0 defining their equilibrium positions. The conventional microwave resonance experiment is

Fig. 1 Schematic of a microwave resonance absorption experiment.

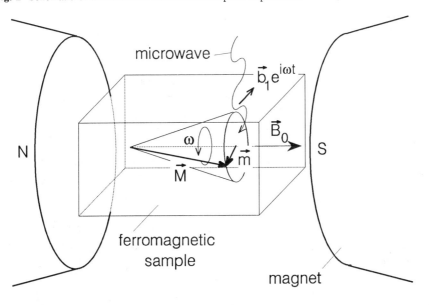

therefore performed by exposing the sample simultaneously to a static magnetic field B_0 and to the high-frequency microwave field b (Fig. 1).

The interaction of the external field with either a spin or orbital magnetic moment exerts a torque which leads to a precessional motion of the moment. In a magnetically ordered material such as a ferro- or ferrimagnet the motion of the individual moments is not independent from each other but phase-correlated by the magnetic interactions. As a consequence the equation of motion can be formulated for the total magnetization M:

$$\frac{dM}{dt} = -\gamma(M \times B) + \frac{\alpha}{M}\left(M \times \frac{dM}{dt}\right) \quad (4)$$

with $B = \mu_0 H$ and $\gamma = g\mu_B/\hbar$ which is the gyromagnetic ratio. g is the Landé g-factor which equals 2 when the orbital contribution to the magnetic moment is quenched as is the case in the iron group. The second term phenomenologically takes into account the damping effects and has been introduced in the given form by Gilbert [28]. The magnetic field entering in Eq. (4) comprises both the effective local static field $B_{\it{eff}}$ and the high-frequency magnetic field vector b. In a magnetically ordered material the effective internal field can deviate considerably from the value of the externally applied magnetic field B_0:

$$B_{\it{eff}} = B_0 + B_{dem} + B_{ani} + B_{exch} \quad (5)$$

The magnetic dipole-dipole interaction between the magnetic moments in the lattice gives rise to a static demagnetizing field B_{dem} oriented antiparallel to the macroscopic static magnetization of the sample. In addition, the magnetic dipole interaction also induces a dynamical demagnetizing field which is associated with the precessing transverse component of the magnetization, included in Eq. (4). Via the spin-orbit coupling the electric multipoles of the surrounding lattice ions yield a contribution to the total energy which depends on the orientation of the magnetic moment with respect to the crystal axes. This contribution is taken into consideration by the so-called anisotropy field B_{ani}. The exchange interaction between the magnetic moments in the lattice it taken into account by B_{exch}. In ferrimagnetic oxides the exchange interaction is of short range nature and is achieved via superexchange by the outer electrons of the oxygen ions.

In absence of the high-frequency field the magnetization M_0 performs a precessional motion around the axis of the local field at the Larmor frequency

$$\omega_r = \gamma B_{\it{eff}}. \quad (6)$$

In magnetic resonance experiments a high-frequency field $b \exp(i\omega t)$ with a small amplitude b is superimposed on the external field. The high-frequency field gives rise to an additional high-frequency magnetization m. Considering only linear response m obeys the equation of motion

$$i\omega m = \gamma(M_0 \times b + m \times B_0) + \text{damping term} \quad (7)$$

From Equation (7) one sees that only a high-frequency field b which is oriented perpendicular to the static field and to the magnetization can produce a high-frequency response. Taking into account the damping (Eq. 4) and using the definition of Eqs. (1) and (2) for m and b in an infinite isotropic sample the following relations are obtained for the real and imaginary parts of the high-frequency susceptibility $\tilde{\chi}$:

$$\tilde{\chi} = \begin{pmatrix} \chi_{xx} & \chi_{xy} & 0 \\ \chi_{yx} & \chi_{yy} & 0 \\ 0 & 0 & 0 \end{pmatrix} \quad (8)$$

where

$$\chi_{xx} = \chi_{yy} = \frac{(\omega_r + i\omega\alpha)\omega_M}{(\omega_r + i\omega\alpha)^2 - \omega^2} \quad (9)$$

$$\chi_{xy} = -\chi_{yx} = \frac{-\omega\omega_M}{(\omega_r + i\omega\alpha)^2 - \omega^2} \quad (10)$$

with $\omega_M = \gamma\mu_0 M_0$. The high-frequency susceptibility varies with the static external magnetic field B_0 and can be measured either at a constant static magnetic field as a function of the frequency of the microwaves or at a constant microwave frequency as a function of the static external field. The latter condition is normally realized in commonly used spectrometers taking advantage of the enhancement of the microwave field intensity inside a resonant cavity. Fig. 2 schematically shows the field dependence of the real and imaginary parts of the susceptibility at constant microwave frequency.

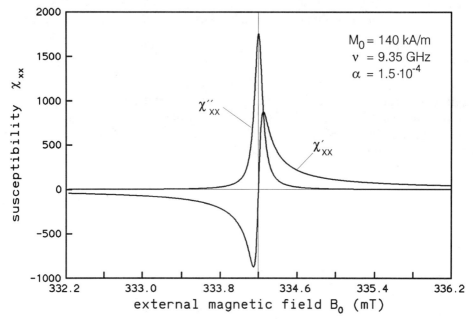

Fig. 2 Magnetic field dependence of the real and the imaginary parts of the high-frequency susceptibility.

The absorption curve $\chi''(\omega)$ has its maximum at the resonance field corresponding to ω_r and a half-width due to the damping $\Delta B = 2\alpha\omega_r/\gamma$.

The resonance absorption of microwaves by the magnetic moments of a ferromagnetic sample is commonly named ferromagnetic resonance (FMR). In ferrimagnetic materials with two magnetic sublattices two ferrimagnetic resonances exist corresponding to the precession of the two sublattice moments in the same or opposite sense. The motion with the sublattice magnetizations precessing in the same sense gives rise to a resonance in the microwave wavelength region and represents the resonance of the net ferromagnetic moment. This ferromagnetic resonance (FMR) in the ferrimagnetic sample is the only one observed in the experiments considered in this chapter. The motion with the sublattice magnetizations precessing in the opposite sense leads to a resonance absorption at much higer energies only accessible to radiations in the infrared wavelength region. In magnetically ordered ferro- and ferrimagnetic compounds the magnetic moments experience an internal field which is superposed on the external magnetic field. In paramagnetic materials the resonance condition for the so-called electron paramagnetic resonance (EPR) is governed solely by the applied external magnetic field.

2. Detection Schemes

The conventional and the main photothermal experimental schemes to measure the high-frequency susceptibility are sketched in Fig. 3. All techniques discussed here make use of a microwave cavity with the sample positioned in a region of maximum magnetic field intensity. Microwaves of constant frequency having the intensity I_0 are supplied by a klystron. By changing the external magnetic field the susceptibility is scanned through its resonance. The resonance absorption is monitored in the *conventional detection* technique by observing the microwave power reflected from the cavity containing the sample. At resonance microwave energy is dissipated in the sample reducing the quality factor Q of the loaded cavity. As a consequence the reflected microwave power is increased.

All photothermal microwave detection techniques are based on the generation of oscillatory thermal waves in the specimen. When applying a modulated heating process a temperature field consisting of an ac-part $\theta(r,t)$ — oscillatory in space and time — and a dc temperature rise is produced in the sample. The spatial variation of $\theta(r,t)$ is governed by the heat diffusion

Fig. 3 Schematic representation of the conventional and of the photothermal detection methods in microwave resonance absorption.

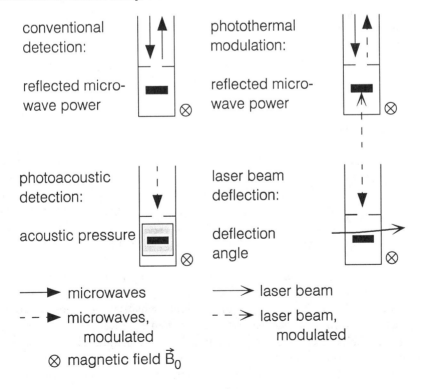

equation [29]:

$$\alpha \nabla^2 \theta(r) - i\omega_m \theta(r) = Q(r) . \qquad (11)$$

Q represents the heat source. The thermal diffusivity α is a function of the mass density ρ, the specific heat capacity c and the heat conductivity κ: $\alpha = \kappa/\rho c$. The solution of the diffusion differential equation (11) is a strongly damped wave with an attenuation length $\mu = (2\alpha/\omega_m)^{1/2}$ which decreases with increasing modulation frequency ω_m. The thermal diffusion length μ also defines the wavenumber of the thermal wave: $k = 1/\mu$ and its phase velocity which increases with increasing moduation frequency: $\upsilon_p = \omega/k = (2\alpha\omega_m)^{1/2}$. Thus, the thermal waves are characterized by a strong dispersion of the amplitude and of the phase velocity.

The *photoacoustic detection* of FMR (PA-FMR) is based on the microwave energy dissipated in the sample during the resonance process. From the sample which is placed in a gas tight chamber heat flows into the gas which itself responds by an increase of the gas pressure [30]. Modulation of the incoming microwave power at audio frequencies leads to periodical pressure fluctuations which can be detected by a commerical microphone. Besides the possibility of narrow-band detection with a lock-in amplifier the modulation of the signal brings about the most significant advantage inherent to photothermal methods which is the depth sensitivity. Changing the thermal diffusion length via the modulation frequency heat generated at different depths below the surface can be monitored.

Local discrimination in addition to the depth dependence is provided by the detection of the thermal wave with the *photothermal laser beam deflection* (PD-FMR). This technique probes the deflection of a laser beam in the optical density gradient caused by the thermal wave in the air adjacent to the sample or in the sample itself [31]. The thermal waves are generated in the same manner as was discussed before. By scanning the probe beam across the sample a lateral image of the microwave absorption density can be obtained. The depth dependence can be measured simultaneously by changing the modulation frequency. The laser beam deflection does not require an air tight sample cell but the resonator has to provide optical access for the probe beam. As we shall see later this constructional feature imposes severe restrictions on the sensitivity and on the spatial resolution of this method.

The *photothermally modulated microwave resonance absorption* (PM-FMR) provides a technique which combines the ability for spatial resolution inherent to the thermal waves and the high sensitivity of the electronic detection used in the conventional experiments [24,26]. An intensity modulated laser or light beam is focused onto the sample and generates a thermal wave inside the specimen. As a consequence of the local temperature change, the resonance absorption of the microwave is modified, which gives rise to a reflected microwave power oscillating coherently with the modulation cycle. As the modulation frequency also controls the depth of the thermal wave penetration into the solid, the recording as a function of modulation frequency and laser beam position can deliver a real three-dimensional tomographic image of the microwave absorptivity [32]. The main difference compared to the other photothermal techniques arises from the fact that the photothermally modulated signal is directly proportional to the temperature derivative of the imaginary part of the high-frequency susceptibility.

3. Experimental Arrangements

All experiments to be described in this chapter have been carried out with a modified electron paramagnetic resonance (EPR) spectrometer working at X-band frequencies around 9.2 GHz. Rectangular and cylindrical resonant microwave cavities are unavoidable for the conventional and the photothermally modulated microwave detection as in these experiments the quality factor of the cavity is the surveyed quantity. The thermal wave techniques based on the photoacoustic effect or on the laser beam deflection monitor the heat released by the sample and therefore can work without a cavity. But there are at least two reasons to utilize a cavity, too. At first, placing the sample in a cavity the enhancement of the microwave power

Fig. 4 Block diagram of the experimental arrangement for the conventional detection and the photothermally modulated FMR experiments.

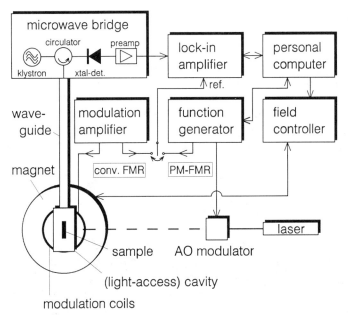

density in the cavity improves the signal to noise ratio. Secondly, simultaneous conventional and photothermal measurements can be performed. Such combined experiments permit an absolute calibration of the photothermal response with reference to the absorbed microwave power density [33].

A Varian E-line EPR spectrometer with an X-band microwave bridge forms the central unit of the experimental arrangements when realizing the four detection schemes of Fig. 3. The microwaves around 9.2 GHz are supplied by a klystron. The signal is always measured at constant microwave frequency as a function of the external static magnetic field. For the photothermal-based detection the input and the output channel of the spectrometer had to be modified.

In the *conventional detection* a bridge technique is used to measure the microwave power reflected at the loaded cavity in resonance (see Fig. 4). The quality factor Q of commercial cavities commonly used in an EPR experiments is of the order of several thousand. To suppress the broad-band noise of the detector diode the resonance is modulated at a fixed frequency (100 kHz) by a supplementary magnetic field generated with Helmholtz coils. The modulated signal is analyzed by a narrow band lock-in amplifier.

For the detection of the *photothermally modulated* microwave response, the conventional signal is branched off from the microwave bridge after the detector and the preamplifier and analyzed synchronously at the frequency of the light beam modulation (see Fig. 4). To provide an optical access for a He-Ne laser (10 - 20 mW) beam a wall-less microwave cavity has to be used. The quality factor of a cavity with an optical window is generally two to three times worse than that of a conventional X-band cavity. In the experiments reported here a wall-less cylinder cavity had been constructed, which is composed of equally spaced concentric annular plates [24,34]. The laser beam being intensity-modulated by an acousto-optic modulator excites inside the sample a thermal wave of a penetration depth that varies with the modulation frequency according to Eq. (3). Thus, the magnetic field modulation used in the conventional detection is replaced by a temperature modulation in the photothermally modulated resonance absorption technique which is local and depth selective. In order to move the modulation spot on the sample either the waveguide arrangement with sample or the exciting laser is mounted on a translational stage which in the reported experiments had a step resolution of 2.5 μm.

The detection schemes based on the photoacoustic effect and on the laser beam deflection measure the direct response of the thermal wave that is excited by the microwave absorption in the sample (Fig. 5). To provide a periodically oscillating microwave power input with sufficient intensity the microwave radiation from the klystron is modulated by a PIN diode and then fed into a travelling wave tube amplifier (not shown in Fig. 5) which in the discussed experiments can provide an output power up to 20 W. The

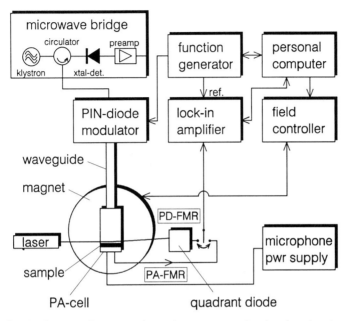

Fig. 5 Block diagram of the experimental arrangement for the detection based on the photoacoustic effect and on the laser beam deflection.

Fig. 6 Shorted end of a waveguide with an adapted photoacoustic cell.

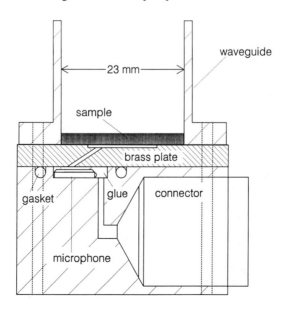

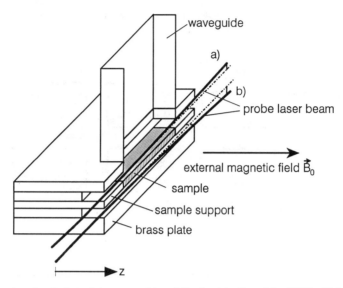

Fig. 7 Cross-sectional view of the waveguide cell for the detection of the FMR with laser beam deflection.

detection units are individually constructed for both methods.

The *photoacoustic effect* requires an air tight chamber with a microphone in gas contact to the sample. Whereas the sample has to be exposed to the maximum microwave power, the microphone has to be protected from the microwave field. To solve this problem two principal designs have been developed. In the first one a quartz tube cell originally developed for EPR measurements to fit in a commercial cavity contains the sample at one end and the microphone at the other end [25,33,35]. This cell takes advantage of the high Q-value of standard cavities and also requires only minor modifications of the spectrometer. A main drawback is the limitation of the sample size defined by the diameter of the quartz tube which is 1.8 mm. For the investigation of larger samples such as the ferrite slabs discussed in the next section the use of a shorted waveguide is more appropriate. Fig. 6 shows the waveguide cell constructed for the PA-FMR studies of flat samples [33,36].

The photoacoustic cell forms one wall of the rectangular cavity. The sample is positioned at a point of maximum magnetic field strength and isolated acoustically from the waveguide by a cover plate made out of glass or lucite. The chamber is in gas contact with the microphone via a short duct which is beyond cutoff for the microwaves, so that no electrical problem concerning the detection electronics arises. As compared to the quartz tube cell the gas volume can be reduced by a factor of 4.5 thus enhancing the PA signal. However, the Q-value of the cavity with the waveguide cell is about 3 times smaller than that of a commercial cavity.

For the laterally resolved measurements using *laser beam deflection* the probe laser beam has to be scanned across the sample in the waveguide. This technical problem can be solved most easily by using a shorted waveguide as shown in Fig. 7.

Two slits are provided to admit the probe beam to pass on the upper and the lower faces of the flat ferrite sample which allows measurements of thermal waves travelling in both directions with respect to the microwave input. To sense the periodical temperature oscillations in the air layer in contact with the sample surface, the vertical deflection of the probe beam is measured by a position sensitive photodiode. To allow lateral scans either the waveguide with the cavity at the bottom is moved with respect to the probe beam, or the probe laser and the detector diode are mounted on a translation stage and the cavity with the sample remains fixed.

4. Sensitivity and Spatial Resolution

The most important merit of the photothermal detection schemes in microwave spectroscopy is the ability of spatial resolution which in general is counterbalanced by a loss of sensitivity. In some cases, however, the photothermal detection can achieve a sensitivity superior to that of the conventional method. Preferential candidates for a photothermal microwave resonance spectroscopy are solids with broad line absorption spectra and samples with a very low density of absorbing centers like a gas of paramagnetic molecules. The inferiority of the conventional detection for broad lines results from the vanishing response on the magnetic field modulation required for the lock-in detection of the electronic signal. In the case of a very weak absorbance the signal of the sample is generally obscured by the background signal due to the absorption in the cavity walls. The photothermal response, however, in general does not experience this background contribution as only the heat emerging from the sample is recorded. Using the photoacoustic method the paramagnetic resonance absorption of oxygen molecules in air could be easily discriminated from that of a solid sample due to the different behavior of the signal intensity when the modulation frequency was changed [35].

The signal to noise ratio and the spatial resolution for the same photothermal detection technique, in general, strongly depend on the system under investigation and the peculiarities of the experimental setup. Therefore, the following discussion of the sensitivity, noise and spatial resolution will be confined to the investigations of the ferrite samples described in detail in the subsequent paragraphs. We shall elaborate quantitative numbers which allow to value the different photothermal methods in comparison with the conventional detection.

4.1 Sensitivity and Signal to Noise Ratio

The intensities of the conventional and of the photothermal signals are governed by the microwave power density absorbed by the sample.

In the case of a matched cavity the microwave power P absorbed by the sample is related to the imaginary part of the susceptibility by [37]

$$\sqrt{P} = \frac{1}{2} Q_{u,0} \eta \chi'' \sqrt{P_0} \qquad (12)$$

where $Q_{u,0}$ is the unloaded Q-value, η is the filling factor of the cavity and P_0 is the power of the incident microwaves of fixed frequency ω. The absorbed power is directly proportional to the imaginary part of the susceptibility χ'' as defined in section 1. For the ferrite materials of Table 2.1 χ'' at resonance has values much larger than unity.

The *conventional detection* is based on the measurement of the microwave power reflected at the cavity due to the change of its quality factor in the course of the resonance absorption [38]. The reflected microwaves are detected by a diode that provides an output voltage ΔU that is proportional to the square root of the reflected power:

$$\Delta U = \frac{1}{2} Q_{u,0} \eta \chi'' \sqrt{Z_0 P_0} \qquad (13)$$

where Z_0 is the impedance of the waveguide connected to the diode. The noise level is essentially controlled by the thermal noise of the detector with an equivalent noise voltage ΔU_{noise} for a given band-width Δf determined by the Nyquist formula:

$$\Delta U_{noise} = \sqrt{4 k_B T Z_0 \Delta f} \qquad (14)$$

where T is the temperature of the diode on the Kelvin scale. The signal to noise ratio of the conventional detection, SNR_{conv}, is then given by the relation

$$SNR_{conv} = \frac{\Delta U}{\Delta U_{noise}} = \frac{1}{2} Q_{u,0} \eta \chi'' \sqrt{\frac{P_0}{4 k_B T \Delta f}} \qquad (15)$$

Thus, the signal to noise ratio of the conventional detection increases with the square root of the input microwave power in contrast to the pure thermal response detected by the photoacoustic effect or by the laser beam deflection which is a linear function of P_0 (Eq. (17) and Sect. 3.2). Inserting typical values of experimental parameters into Eq. (15) one obtains for the described experiments on ferrites at 100 kHz magnetic field modulation and a time constant of the lock-in amplifier corresponding to $\Delta f = 1$ Hz a signal to noise ratio SNR_{conv} of

$$SNR_{conv} \approx \begin{cases} 3 \times 10^8 \quad \text{for YIG} \\ (Q_{u,0} = 7000, \eta = 8 \times 10^{-7}, \chi'' = 1750, P_0 = 50 \mu W) \\ \\ 2 \times 10^7 \quad \text{for Ni-Ferrite} \\ (Q_{u,0} = 1, \eta = 2 \times 10^{-3}, \chi'' = 6, P_0 = 200 mW) \end{cases} \quad (16)$$

The *photoacoustic signal* is proportional to the pressure fluctuations δp in the gas volume surrounding the sample. The amplitude of the acoustic pressure variation is a linear function of the amplitude of the surface temperature oscillations θ_s. For a thermally thick ferrite sample (thermal diffusion length $\mu_s \leq$ geometrical sample thickness d) and in a one-dimensional approximation, which is adequate for disk or slab-shaped samples, the pressure and temperature amplitudes are given by the equations

$$\theta_s = \frac{P}{A_s \varepsilon_s \sqrt{\omega_M}} \quad (17)$$

and

$$\delta_p = (\gamma - 1) \frac{A_s}{V_g} \theta_s \frac{\varepsilon_g}{\sqrt{\omega_M}} \quad (18)$$

where P is the absorbed microwave power according to Eq. (12). γ represents the coefficient of adiabatic expansion. A_s and V_g are the surface of the sample and the volume of the gas cell. ω_M is the frequency of modulation of the microwave input power. $\varepsilon = \sqrt{\kappa \rho c}$ is the thermal effusivity of the gas (index g) and of the sample (index s) which reflects the ability of a material to accept an incident heat flow. Whereas the thermal diffusivity $\alpha = \kappa/\rho c$ (see Sect. 2) determines the heat flow inside the sample or in the surrounding gas, the effusivity ε governs the amount of heat transferred into the matter by the thermal contact.

The detection limit of the photoacoustic sensor is determined by the thermal fluctuations of the gas and by the electronic noise of the microphone. Theoretical expressions for the thermal pressure fluctuations have been derived by Kreuzer for resonant photoacoustic cells [39]. Inserting the characteristics of the photoacoustic cells used in the present experiments one obtains a value of δp_{fluct} = 1 μPa that corresponds to an effective surface temperature amplitude of $\theta_{s,fluct}$ = 10^{-7}K. this value is at least 3 times less than the experimentally observed noise level. From this result we conclude that the dominant part of the noise is due to the microphone detector channel.

In order to be able to compare the figure of merit of the different detection schemes we shall refer to numbers deduced at the same defined conditions: thermally thick ferrite slab, microwave frequency 9.2 GHz, microwave power P_0 = 200 mW, $Q_{u,0}$ = 1, modulation frequency f_m = 300 Hz. For these conditions the amplitude of the photoacoustic signal is typically δp = 0.45 Pa. With a measured equivalent noise pressure amplitude of typically δp_{noise} = 5 × 10^{-6} Pa the signal-to-noise ratio provided by the photoacoustic detection is $SNR_{PA} \approx 2 \times 10^5$ which is about two orders of magnitude less than the S/N ratio achieved with the conventional technique. Both signal and noise pressure depend on the modulation frequency with an increasing signal- to-noise ratio as the modulation frequency is lowered [25]. Based on the same physical origin at a constant modulation frequency the S/N ratio can be considerably increased if the sample thickness is reduced. A detailed theoretical and experimental study of the photoacoustic signal from microwave resonance absorption in paramagnetic spheres as a function of the sphere size proved that below a critical sphere size the photoacoustic signal compares with that of the conventional recording [40-42]. This observation can be easily understood, as for a sample with a size smaller than the thermal diffusion length practically all the heat generated in the sample can be transferred to the gas. Therefore, in addition to the situations discussed at the beginning of this section, thermal wave methods gain much attraction as compared to the conventional microwave detection techniques when the samples are thermally thin.

We already mentioned that an estimate of the signal to noise ratio for the photothermal detection schemes depends on the details of the experiment. This is particularly true for the detection of the microwave resonance absorption using *photothermal deflection (PD)*. In a microwave experiment the high sensitivity and the lateral resolution of the laser beam deflection [43] suffer from restrictions imposed above all by the use of a cavity and the necessity of applying a large external magnetic field. The cavity hampers the entrance of the probe beam (Fig. 7) and the large sizes of both the cavity and of the magnet require long focus optical arrangements. In the experiments described in the next sections under the same conditions the equivalent noise temperature amplitude $\theta_{PD,noise} \approx 10^{-5}K$, which is about two orders of magnitude higher than that of the photoacoustic measurement. A large portion of this noise level is due to the mechanical instability of the detector system owing to the large geometrical dimension e.g. the spatial separation of the probe laser, the sample, and the detector. The signal amplitude at given standard conditions was typically $10^{-3}K$ leading to a S/N ratio of $SNR_{PD} = 10^2$ which, again, is two orders less than that of the photoacoustic detection. In order to improve the sensitivity of the PD-detection we used a TWT amplifier that provided an enhanced input power by at least one order of magnitude. In addition, one may treat the signal by filtering techniques to reduce low frequency noise. This procedure can improve the signal-to-noise ratio by at least one order of magnitude as it was shown recently by Woelker et al. for the case of a beam deflection experiment using laser excitation [44].

The most promising photothermal method with respect to sensitivity and spatial resolution is the *photothermally modulated microwave absorption*. This method is based on the local temperature modulation of the resonance condition achieved by the heating with a periodically modulated laser beam [24]. The response to this local heating is measured by the conventional technique coherently with the modulation cycle of the laser heating. Thus, the photothermally modulated resonance signal is essentially the temperature derivative of the conventional signal, Eq. (13), spatially weighted with the amplitude $\theta_s(r)$ of the thermal wave generated in the sample of volume V_s. The amplitude ΔU_{PM} of the oscillatory voltage at the detector diode is then given by the equation:

$$\Delta U_{PM} = \frac{1}{2} Q_{u,0} \eta \sqrt{Z_0 P_0} \frac{1}{V_s} \int_{V_s} \left(\frac{\partial \chi''}{\partial T} \theta_s \right) dV \qquad (19)$$

In deriving Equation (19) it has been assumed that the microwave power density is constant in the volume region that is reached by the thermal wave.

For the experiments described here the noise level was essentially controlled by the thermal noise of the detector. The equivalent noise voltage ΔU_{noise} is therefore the same as for the conventional detection. Using the Eqs. (14) and (19) one obtains the following expression for the signal-to-noise ratio for the photothermally modulated microwave resonance absorption:

$$SNR_{PM} = \frac{\Delta U_{PM}}{\Delta U_{noise}} = \frac{1}{2} Q_{u,0} \eta \frac{P_0}{\sqrt{4 k_B T \Delta f}} \frac{1}{V_s} \int_{V_s} \left(\frac{\partial \chi''}{\partial T} \theta_s \right) dV \qquad (20)$$

This is nearly the same relation as for the conventional detection only the imaginary part of the high frequency susceptibility is replaced by the spatially weighted average of the temperature derivative of the susceptibility. For typical parameters used in our experiments with a cylindrical TE_{011} light access cavity ($Q_{u,0} \approx 3000$) and a 1 mW laser beam for the excitation of the thermal wave at a modulation frequency of 300 Hz one obtains as the lower detectable limit of the temperature derivative of the high frequency susceptibility $\partial \chi''/\partial T \approx 10^{-5} K^{-1}$. All other parameters have been the same as adopted for the conventional detection. For YIG at room temperature $\partial\chi''/\partial T = 10^{-3} K^{-1}$.

From Eqs. (19) and (20) we see that the signal of the photothermally modulated resonance depends on both the magnetic properties represented by the temperature derivative of the susceptibility and on the thermal properties comprised by the thermal wave amplitude θ_s.

An inherent property of the thermal wave is the variation of the amplitude and of the phase shift with the modulation frequency establishing the depth sensitivity of this method. Therefore, the frequency dependence also offers a tool to discriminate between surface and volume absorption of the microwave. For the two limiting cases of a homogeneous volume absorption and of a surface absorption the susceptibility derivative can be extracted from the integral in Eq. (19) if the sample has homogeneous thermal properties thus yielding the following frequency dependence for the PM-FMR signal:

Volume absorption (absorption length $\geq$ thermal diffusion length):

$$\Delta U_{PM} \propto \omega^{-1} \qquad (21a)$$

Surface absorption (absorption length $\leq$ thermal diffusion length):

$$\Delta U_{PM} \propto \omega^{-1/2} \qquad (21b)$$

The theoretically expected behavior is very well reproduced by the experimental data. In Fig. 8 the amplitude and phase angle of the

Fig. 8 Modulation frequency dependence of the photothermally modulated FMR signal amplitude of the first volume mode in a YIG film [32].

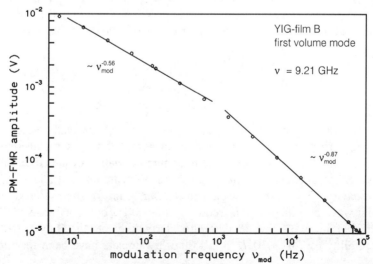

photothermally modulated signal from a YIG film are shown as a function of the modulation frequency [32]. For lower frequencies the film acts as a surface absorber on the substrate and according to Eq. (21) a one over square root dependence for the modulation frequency is observed. At very high frequencies a cross-over to the situation according to Eq. (20) becomes noticeable as at these frequencies the film thickness and the thermal diffusion length are of the same order of magnitude.

The magnetic properties on the photothermally modulated signal enter through the temperature derivative of the imaginary part of the high frequency susceptibility $\partial \chi''/\partial T$. Generally, the temperature dependence of the susceptibility has several sources. Parameters affecting the high-frequency susceptibility are the temperature dependence of the spontaneous magnetization M_s, the magnetocrystalline anisotropy K_1, K_2, the magnetoelastic coupling constant λ and the spin-lattice relaxation time τ:

$$\chi''(T) = \chi''[M_s(T), K_i(T), \lambda(T), \tau(T), \ldots] \quad (22)$$

In magnetically ordered materials such as the ferrites the most prominent contribution to $\partial \chi''/\partial T$ arises from the temperature dependence of the magnitude and in some special cases of the orientation of the spontaneous magnetization M_s [45]. Owing to the nonlinear temperature dependence of the magnetization, the photothermally modulated signal intensity varies

Fig. 9 Temperature dependence of the spontaneous magnetization in a *ferro*magnet obeying a $B_{5/2}$ Brillouin function and in the *ferri*magnetically ordered material (YIG) resulting from a superposition of two *differently* weighted $B_{5/2}$ Brillouin functions.

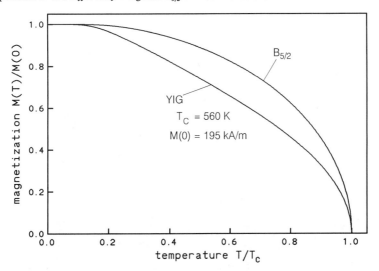

strongly with temperature. In a ferromagnet the spontaneous magnetization follows a Brillouin function sketched in Fig. 9. In a ferrimagnetic material like YIG due to the presence of two sublattices the magnetization is described by a linear superposition of two Brillouin functions [46].

A PM-FMR signal typical for a signal generation process governed by the temperature dependence of the magnitude of the spontaneous magnetization is shown in Fig. 10. For an experimental configuration where the external magnetic field is oriented parallel to the sample surface the peak of the microwave resonance absorption if shifted to higher magnetic fields as the magnetization is lowered in magnitude. Simultaneously, the intensity of the resonance absorption is reduced leading to the asymmetry displayed by the PM-FMR signal from the YIG film. The signals were recorded at a microwave frequency of 9.2 GHz using a modulation frequency of the laser beam of 25 kHz.

A distinct temperature range exists near the Curie temperature T_c where the theoretical value of the temperature coefficient of the magnetization diverges, leading to a strong enhancement of the PM-FMR signal. In real crystals this divergence is smeared out by short range ordering of the magnetic moments already above T_c. As compared to a pure ferromagnet the magnetization of ferrimagnetic materials exhibits a more complicated temperature behavior (Fig. 9) due to the competing superposition of at least two sublattice spontaneous magnetizations. However, approaching the ferrimagnetic Curie temperature from below an enhancement of the PM-FMR

Fig. 10 Magnetic field dependence of the amplitude and of the phase of the photothermally modulated ferrimagnetic resonance signal from a YIG film.

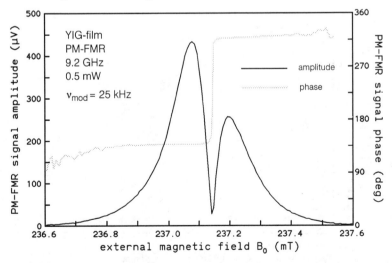

signal amplitude is also expected in a ferrimagnet. We have performed a detailed study of the temperature dependence of the PM-FMR signal in yttrium iron garnet (YIG) [47]. Results are shown in Fig. 11. The theoretical curves have been calculated taking into account the measured temperature variation of the spontaneous magnetization only. Within the scope of this assumption, the variation of the signal intensities of the conventional and the photothermally modulated signal are well described in the temperature range above 250 K. The deviations at lower temperatures can be qualitatively explained by the broadening of the resonance linewidth due to paramagnetic impurities and temperature induced changes of thermal parameters [32]. Particularly, the decrease of the specific heat leads to a strong enhancement of the PM-FMR response compared to the conventional signal. Therefore, towards lower temperatures the ratio of the PM-FMR signal to the conventional signal can increase, although the $\partial \chi''/\partial T$ becomes smaller.

4.2 Spatial Resolution

The photothermal detection schemes are the only methods that can provide a depth dependent spatial resolution of the microwave resonance

Fig. 11 Experimental and theoretical temperature dependence of the FMR and the PM-FMR signals in YIG.

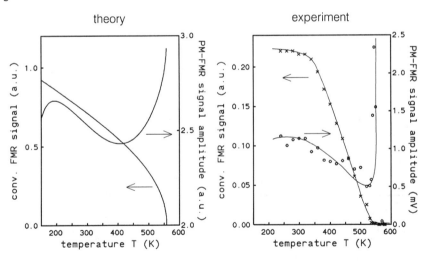

absorption on the micrometer scale. The resolution is determined by the thermal diffusion length and for the laser assisted techniques by the laser beam diameter and the size of laser spot, respectively.

The *photoacoustically detected* microwave resonance offers spatial discrimination via the penetration depth of the thermal wave. This depth resolution is therefore governed by the modulation frequency as the thermal diffusion length decreases with the inverse square root of ω_m (see Sect. 2). For a particular material the depth resolution is limited by the highest frequency still detectable by the microphone. The numbers given in Table 2 are representative for ferrite materials. Metals which are good thermal conductors have larger μ_{min} values and for plastic materials, which possess much smaller thermal conductivities, lower scale spatial resolution may be accessible.

Table 2 Experimental parameters for the photothermal detection of ferrimagnetic microwave absorption. The values given for the PA technique result from experiments with soft ferrite slabs (Sect. 3.5). The data for the PD-FMR represent results for soft ferrites (Sect. 3.6) and YIG (Sect. 4.3), whereas the PM technique numbers are only given for YIG (Sect. 4.4). The thermal parameters for a typical soft ferrite are $\kappa = 3.9$ Wm^{-1}K^{-1}, $\rho = 4000$ kg/m^3 and $c = 925$ Jkg^{-1}K^{-1}. The YIG data are $\kappa = 7.4$ Wm^{-1}K^{-1}, $\rho = 5170$ kg/m^3 and $c = 25.5$ Jkg^{-1}K^{-1}.

	PA Photoacoustic Detection	Large PD Photothermal Beam Deflection	PM Photothermal Modulation
SENSITIVITY photothermal/ conventional: θ_{min}:	$10^{-3} - 10^{-1}$ 10^{-7} K	$10^{-3} - 10^{-5}$ 10^{-5} K	$10^{-1} - 10^{1}$ 10^{-8} K
DEPTH RESOLUTION f_{max}: μ_{min}:	7 kHz[a] 6.9 μm	30 kHz[a]/70 kHz[b] 3.3 μm/16 μm	5 MHz[b] 2 μm
LATERAL RESOLUTION	no -	1-dimensional 300 μm	2-dimensional 20 μm diam.

[a] soft ferrites   [b] YIG

The *photothermal beam deflection* technique principally offers a depth resolution superior to that of the photoacoustic detection. The achievable resolution markedly depends on the experimental conditions for a beam deflection experiment in a waveguide or cavity. In a previous PD-FMR study of magnetic tapes the signal-to-noise ratio restricted the upper limit of the modulation frequency to about 15 kHz [33, 48], whereas in the PD experiments with ferrite samples frequencies up to 70 kHz could be achieved. The PD-FMR is the only photothermal method that offers a lateral resolution based on the direct detection of the thermal wave. The area probed by the focused laser beam in grazing incidence on the sample is elongated along the laser beam and confined in the perpendicular direction. Therefore, we consider the lateral resolving power of the PD-FMR as essentially one-dimensional. The values given in Table 2 are representative for the experiments described in this article.

The most promising results with respect to spatial resolution are provided by the *photothermally modulated* resonance absorption. Due to the electronic detection the upper limit of the modulation frequency can be as high as a few MHz leading to depth resolutions in the submicron range. As the laser exciting the thermal wave can be incident perpendicular to the sample surface, a much smaller spot size can be reached as in the case of the photothermal deflection technique. Again the data shown in Table 2 are based on the results obtained from the samples discussed later in this chapter.

III. APPLICATIONS TO FERRIMAGNETIC MATERIALS

In this Section we outline the application of photothermal microwave techniques to the investigation of ferrimagnetic materials which are of great interest for most of today's microwave applications. These materials are shortly introduced in section III.1. Although these materials are magnetically homogeneous certain excitations of the spin system resulting in a nonuniform distribution of energy inside the sample can be induced. These excitations depend on different parameters, of which sample size and energy of the incident radiation are the most important. Depending on the main contributions to the dispersion relation, these two kinds of modes are called magnetostatic modes (dipolar coupled spin-waves) or spin-wave modes (exchange coupled spin-waves).

In a first step we treat simple electromagnetic propagation in a system in which under the present experimental conditions spin-wave can be neglected. Technical ferrites are placed in front of a short in a waveguide so that due to multiple reflections of the incident microwaves a spatially inhomogeneous power density distribution is created in the specimen which is well-known in the literature [49]. This distribution markedly differs from the often used special cases of surface or bulk absorption. The description of this problem from the thermal and electromagnetic point of view is given in Sect. III.3. After showing the abilities of photothermal microwave methods magnetostatic modes in yttrium iron garnet slabs and spheres are dealt with in Sect. III.4. In these samples the lateral inhomogeneity in the absorption of power is more pronounced than in the ferrites treated before. In addition to providing an ideal material for testing photothermal detection schemes, the investigation of magnetostatic modes in low-loss ferrimagnetic materials such as yttrium iron garnet results in new experimental evidence not available to any other experimental technique.

1. Magnetic Properties of Ferrimagnetic Materials

Ferrites and garnets are the most prominent ferrimagnetic materials which are composed of metal oxides with the iron oxide Fe_2O_3 as the principal constituent. The ferrites are derivates of magnetite with the general formula MO, Fe_2O_3 where M is a divalent metal ion such as Mn, Mg, Zn or Ni and crystallize in the cubic face centered spinel structure [50]. The garnets contain only oxides of trivalent metals. The most prominent member of this class is the yttrium iron garnet (YIG). The cubic lattice of YIG contains eight formula units in the cubic unit cell and has the chemical formula $3Y_2O_3$, $5Fe_2O_3$ equivalent to $Y_3Fe_5O_{12}$ [51, 46]. Frequently mixed garnets are grown where yttrium is partially substituted for by other rare earths or by gallium. A third class is made up by the hexagonal ferrites which crystallographically belong to the iron plumbites. The best know member is barium ferrite $BaFe_{12}O_{19}$ which nowadays has received renewed attention as base material for the fabrication of particulate magnetic tapes [52].

At the ferrimagnetic Curie temperature T_C ferrites and garnets undergo a magnetic transition to a ferrimagnetic order. In the ferrimagnetic state the magnetic moments of two or more sublattices are oriented antiparallel. But in contrast to an antiferromagnet the magnitudes of the sublattice magnetizations are not equal in a ferrimagnet giving rise to a resultant magnetic moment in the ordered state. Therefore, ferrimagnets behave in a similar fashion to ferromagnetic materials with a lower net spontaneous magnetization. But as the exchange interaction is large, the ferrimagnetic Curie temperature T_C is comparable with those of strong ferromagnets. The competition between the different sublattice magnetizations may also give rise to a peculiar temperature dependence of the resultant magnetization with a compensation temperature T_K. At T_K the spontaneous magnetizations of the sublattices have the same magnitude but opposite sign, so that the resultant magnetization vanishes. Some magnetic properties of selected ferrites and garnets are collected in Table 3. As compared to the cubic ferrites, on which we report in this chapter only, the hexaferrites are distinguished by their large uniaxial anisotropy field meeting one important requirement for a permanent magnet.

	Ferrimagnet	T_C/K	T_K/K	M_s/T	B_{an}/mT
soft ferrite	$NiFe_2O_4$	858 [a]	-	3400 [a]	-460 [a]
soft ferrite	$Ni_{0.8}Zn_{0.2}Fe_2O_4$	768 [a]	-	4900 [a]	-150 [a]
garnet	$Y_3Fe_5O_{12}$	550 [b]	-	1760 [b]	-8.7 [c]
garnet	$Gd_3Fe_5O_{12}$	564 [b]	290 [b]	122 [b]	-20.6 [b]
hexaferrite	$BaFe_{12}O_{19}$	725 [a]	-	4800 [a]	1700 [a]

[a] after [53] [b] after [46] [c] after [54]

Table 3 Magnetic properties of ferrites at 20° C: Curie temperature T_C, compensation temperature T_K, saturation magnetization M_s and anisotropy field B_{an}.

The technical importance of the ferrites and garnets is established by their particular high frequency magnetic properties. In an oscillating electromagnetic field the response to the magnetic field component is determined by the motion of the magnetic moments which will be discussed in more detail in the next section. The response to a time dependent electric field is commonly described by the frequency dependent dielectric tensor which is a complex quantity. In the low frequency range from a few Hz to a few hundred Hz the ferrites and garnets behave like strongly lossy dielectrics [55]. With increasing frequency both the real and imaginary part of the dielectric tensor components decrease by several orders of magnitude with the real part approaching a limiting value of about 10 in the GHz region. This good "dielectric transparency" of the ferrites and garnets in the wavelength range of the microwaves provides a very important pre-condition in order to make use of the magneto-optical properties in microwave components. Most of the solids suited from the magnetic point of view for this purpose are metals where the microwaves are prevented to enter into the material by the skin effect.

2. Magnetic Excitations

The properties of the propagation of plane waves in an unbounded medium when taking into account the exchange interaction allow the depiction of the possible excitation characteristics in a ferro- or ferrimagnetic material. This is the basis for the formulation of those equations governing the electromagnetic excitations, the photothermal investigation of which is presented in the following paragraphs.

The treatment of the plane wave solution of the equation of motion in an infinite medium including the "exchange term" $\omega_{ex}a^2k^2$ (Ref. [49], Sect. 4-5) results in the definition of three regions for the wavevector k where different dispersion characteristics are exhibited.

Starting from Maxwell's equations

$$\nabla \times \mathbf{e} = -i\omega\tilde{\mu}\cdot\mathbf{h}, \quad \nabla \times \mathbf{h} = i\omega\tilde{\varepsilon}\mathbf{e} \tag{23}$$

for harmonic time dependence the propagation constant $\Gamma = \alpha + i\beta$ ($= ik$, for the lossless case) of the magnetic field $\mathbf{h} = \mathbf{h}_0 \exp[-\Gamma(\mathbf{n}\cdot\mathbf{r})]$ yields for propagation in the yz-plane [56]:

$$\frac{k^2}{k_m^2} = \frac{(\mu^2 - \mu - \kappa^2)\sin^2\theta_k + 2\mu \pm [(\mu^2 - \mu - \kappa^2)]\sin^4\theta_k\theta_k + 4\kappa^2\cos^2\theta_k]^{\frac{1}{2}}}{2[(\mu - 1)\sin^2\theta_k + 1]} \tag{24}$$

k_m is the wave number for propagation of light in a medium with relative permittivity ε_r, θ_k is the angle between the direction of propagation and the d.c. magnetic field vector and μ and κ are the components of the permeability tensor, Eqs. (3) and (8) which are, still neglecting losses, slightly modified to include the exchange term:

$$\mu = 1 + \frac{\omega_r\omega_M}{\omega_r^2 - \omega^2}, \quad \kappa = \frac{\omega\omega_M}{\omega_r^2 - \omega^2}. \tag{25}$$

ω and ω_M are well-known from Eq. (9). ω_r now includes the exchange term $\omega_r = \gamma B + \omega_{ex}a^2k^2$.

The corresponding electric and magnetic fields are given by [56]:

$$\mathbf{h} = \begin{pmatrix} (k_m/k)^2 \\ (i/\kappa)[\mu(k_m/k)^2 - 1] \\ i\dfrac{[\mu(k_m/k)^2 - 1]}{k} \cdot \dfrac{\cos\theta_k \sin\theta_k}{\sin^2\theta_k - (k_m/k)^2} \end{pmatrix} h_0 e^{-i\mathbf{k}\cdot\mathbf{r}}, \tag{26}$$

$$\mathbf{e} = -\frac{k_m}{k}\left(\frac{\mu_0}{\varepsilon}\right)^{\frac{1}{2}} \begin{pmatrix} i\dfrac{[\mu(k_m/k)^2 - 1]}{\kappa} \cdot \dfrac{\cos\theta_k}{\sin^2\theta_k - (k_m/k)^2} \\ \cos\theta_k \\ -\sin\theta_k \end{pmatrix} h_0 e^{-i\mathbf{k}\cdot\mathbf{r}}. \tag{27}$$

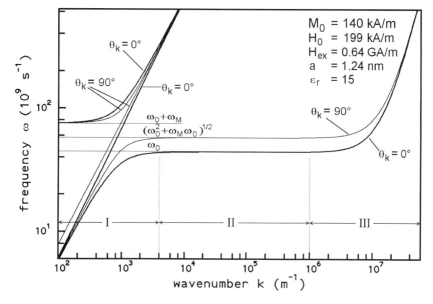

Fig. 12 Dispersion relation for the propagation of plane waves in an infinitely extended ferromagnetic medium. θ_k is the angle between the wavevector k and the external magnetic field H_{ext}. The marked regions are associated with electromagnetic propagation (I), magnetostatic modes (II) and spin-wave-modes (III).

The dispersion relation Eq. (24) for every angle of propagation yields a solution with three branches as shown in Figure 12. The manifold of solutions is bounded by curves for propagation at angles of $\theta_k = 0°$ and $\theta_k = 90°$. The ordinary branch $\omega \propto k$ describes the propagation of electromagnetic plane waves in a dielectric with relative permittivity ε_r. The two extraordinary branches result from the coupling of pure electromagnetic propagation and spin-wave propagation. The branch that shall be discussed here is the lower one. As a function of a normalized wavevector $k' = k/k_m$ the dispersion curve can be divided into three regions where different processes contribute most to the resultant curve.

In region (I), the region of electromagnetic propagation, where $k' \approx 1$

the magnetic and electric fields Eqs. (26),(27) are of the same order of magnitude and only differ slightly from those in a normal dielectric. This region, in absolute values, extends up to $k = 4000$ m^{-1}. This limiting value, however, is not to be taken as a sharp border but one has to allow for an intermediate region where propagation characteristics mix up.

The second region (II) extends from $k = 4000$ m^{-1} to $k = 10^6$ m^{-1}. In this region, the phase velocity being very small and the group velocity being exactly zero, the exchange interaction does not yet influence the dispersion curve. This is because the wavenumber k in the exchange term $\omega_{ex}a^2k^2$ does not yet surpass the product of the other two terms which include the exchange field and the lattice constant. So the condition for the second region has the form $1 \ll k' \ll \omega_0/\omega_{ex}a^2k^2$.

Neglecting terms in $1/k'^2$ in the expressions for the fields, Eqs. (26) and (27), the electric field vanishes while the magnetic field takes on a simple form that results in the condition for magnetostatic propagation:

$$\nabla \times \boldsymbol{H} = 0 . \tag{28}$$

Depending on the direction of propagation the dispersion relation is now given by

$$\omega^2 = \omega_0(\omega_0 + \omega_M \sin\theta_k) . \tag{29}$$

In region (III) beyond 10^6 m^{-1} k is sufficiently large so that the exchange term must not be neglected any more: $\omega_0/\omega_{ex}a^2k^2 \ll k'$. This region is called the region of exchange propagation. The statements made for the electric and magnetic fields in region (II) are also valid here with the exception that ω_0 has to be replaced by $\omega_0 + \omega_{ex}$ so that the dispersion relation

$$\omega^2 = (\omega_0 + \omega_{ex}a^2k^2)(\omega_0 + \omega_{ex}a^2k^2 + \omega_M \sin\theta_k) \tag{30}$$

now has the well known form for the spin-wave manifold.

3. Microwave Density Distribution in Soft Ferrite Slabs

In this section the photothermal investigation of soft ferrite slabs by means of the PA-FMR and the PD-FMR techniques will be discussed. In the experiments described below, different ferrite slabs with thicknesses of 1 - 3 mm and lateral dimensions equalling those of the X-band waveguides used are subjected to microwave radiation in a nonresonant waveguide arrangement. As the samples are placed directly in front of a short in the waveguide, this leads to a spatially inhomogeneous distribution of the microwave power density inside the magnetically homogeneous materials. The power distribution is influenced by the external magnetic field and differs markedly from the commonly assumed surface or volume absorption for photothermal experiments. The two abovementioned absorption patterns are only limiting cases of what can be observed in our experiments. The theory discussed later yields both an exponential decay (where the decay constant indicates surface or volume absorption) and an oscillatory behavior of the power density. In a conventional FMR spectrum this leads to the observation of "body-resonances" [57, 49]. The photothermal signal for the problem of inhomogeneous heating of a homogeneous sample will be derived in the following section. In addition to the depth dependent information already accessible to the PA-FMR experiment, PD-FMR measurements reveal the influence of the not negligible inhomogeneous demagnetizing fields inside the samples.

In the following section III.3.1 the distribution of the electric and magnetic fields inside a ferrite slab in front of a short will be calculated. With this result it is then possible to calculate the overall absorbed microwave power and the depth dependent absorbed power density. In section III.3.2 the PA- and the PD-signals will be derived for an arbitrary absorption distribution in a homogeneous material. The sections III.3.3-III.3.6 deal with experimental results.

3.1 FMR in Thin Ferrite Slabs

The scope of this section is the calculation of the microwave power density P absorbed in a ferrite slab in front of a short in a waveguide. To obtain P we assume a one-dimensional model for a ferrite of thickness d and infinite lateral dimensions. From the solution of the boundary value problem for plane wave electromagnetic propagation one obtains the field distribution in the sample. The propagation constant occurring in these expressions can be deduced from Maxwell's equations together with the equation of motion. P is then calculated from Poynting's theorem.

Fig. 13 The 1-D model for electromagnetic propagation.

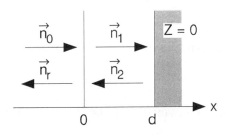

The one-dimensional model for electromagnetic propagation is displayed in Fig. 13. In the free space above the sample ($x < 0$) we assume incident and reflected plane waves with wave vectors $\boldsymbol{k}_i$ and $\boldsymbol{k}_r$, respectively. The ferrite medium extends from $x = 0$ to $x = d$ and supports propagation of waves k_1 and k_2. The backing of the ferrite ($x > d$) is a metal. Therefore, neglecting the skin effect, no waves travel in this region. Z is the boundary impedance.

The wave vectors $\boldsymbol{k}$ can be written as the product of the complex propagation constant Γ and the unit vector $\boldsymbol{n}$. Then, omitting the oscillatory parts, the time independent parts of the plane electric and magnetic waves have the following form:

$$\boldsymbol{e}, \boldsymbol{h} \sim \exp(\boldsymbol{k} \cdot \boldsymbol{r}) = \exp[\Gamma(\boldsymbol{n} \cdot \boldsymbol{r})] \tag{31}$$
$$= \exp[\Gamma(x \sin\phi + z \cos\phi)]$$

From Maxwell's equations for $\nabla \times \boldsymbol{e}$ and $\nabla \times \boldsymbol{h}$ and Eqs. (3) and (8) for the case of transverse magnetization ($\phi = 90°$, $\boldsymbol{n} \perp \boldsymbol{H}_0$) the propagation constant for transverse electric (TE) waves yields

$$\Gamma = i\frac{\omega}{c}\sqrt{\varepsilon_r \mu_r}$$
$$\mu_r = \frac{\mu^2 - \kappa^2}{\mu}. \tag{32}$$

ω is the microwave frequency, $c = 1/\sqrt{\varepsilon_0 \mu_0}$ is the speed of light in free space, ε_r is the relative permittivity in a medium, μ_r is defined above and contains μ and κ the diagonal and nondiagonal components of the permeability tensor (3) and (8). The propagation constant in free space Γ_0 therefore is

$$\Gamma_0 = i\frac{\omega}{c}. \tag{33}$$

As μ_r consists of the complex quantities μ and κ, μ_r itself is also complex:

$$\mu_r = \mu_r' - i\mu_r'' \tag{34}$$

Thus the propagation constant Γ may be divided into real and imaginary parts:

$$\Gamma = \alpha + i\beta,$$

$$\alpha = \frac{\omega}{c} \sqrt{\frac{\varepsilon_r}{2}} \sqrt{|\mu_r| - \mu_r'} \qquad (35)$$

$$\beta = \frac{\omega}{c} \sqrt{\frac{\varepsilon_r}{2}} \sqrt{|\mu_r| + \mu_r'} \qquad (36)$$

With the result from Eqs. (3) and (8) and in the limit of small damping, μ_r can be written as:

$$\mu_r \approx \frac{\omega_0^2(1 + \chi_0)^2 - \omega^2 + 2i\omega\omega_r(1 + \chi_0)}{\omega_0^2(1 + \chi_0) - \omega^2 + 2i\omega\omega_r(1 + \frac{1}{2}\chi_0)} \qquad (37)$$

For our experiments where the microwave frequency is kept constant and the external magnetic field B_0 is varied, we need μ_r as a function of B_0:

$$\mu_r \approx \frac{(B_0 + M_0)^2 - B^2 + 2i\Delta B(B + \frac{B}{B_0} + M_0)}{B_0(B_0 + M_0) - B^2 + 2i\Delta B(B + \frac{B}{2B_0} + M_0)} \qquad (38)$$

B is the inner resonance field given by $\omega = \gamma B$ and ΔB is connected to the relaxation time τ_r by $1/\tau_r = \omega_r = \gamma\Delta B$.

Now we can solve the boundary value problem

vacuum: incident wave:

$$\boldsymbol{e}_i = e_i \exp(\Gamma_0 x)\hat{z}, \quad \boldsymbol{h}_i = \frac{e_i}{Z_0}\exp(\Gamma_0 x)\hat{y} \qquad (39a)$$

reflected wave:

$$\boldsymbol{e}_r = e_r \exp(\Gamma_0 x)\hat{z}, \quad \boldsymbol{h}_r = \frac{e_r}{Z_0}\exp(\Gamma_0 x)\hat{y} \qquad (39b)$$

ferrite: incident wave:

$$e_1 = e_1\exp(\Gamma x)\hat{z}, \quad h_1 = \frac{e_1}{Z}\exp(\Gamma x)\hat{y} \tag{40a}$$

reflected wave:

$$e_2 = e_2\exp(\Gamma x)\hat{z}, \quad h_2 = \frac{e_2}{Z}\exp(\Gamma x)\hat{y} \tag{40b}$$

Z_0 and Z are the impedances in free space and in the ferrite. In the present case

$$Z_0 = e_0/h_0 = -e_r/h_r = \sqrt{\mu_0/\varepsilon_0} \tag{41}$$

and

$$Z = e_1/h_1 = -e_2/h_2 = Z_0\sqrt{\mu_r/\varepsilon_r} . \tag{42}$$

The boundary conditions are the continuity of the tangential components of the fields at $x = 0$ and a short (i.e. impedance $Z = 0$) at $x = d$. Together with Eq. (41) the result for h and e is:

$$h = \begin{pmatrix} -i\dfrac{\kappa}{\mu} \\ 1 \\ 0 \end{pmatrix} h_y, \quad e = \begin{pmatrix} 0 \\ 0 \\ 1 \end{pmatrix} e_z ; \tag{43}$$

$$h_y(x) = \frac{2h_0 Z_0 \cos[\Gamma(d-x)]}{Z_0\cosh(\Gamma d) + iZ\sinh(\Gamma d)},$$

$$e_z(x) = \frac{2ih_0 ZZ_0 \sin[\Gamma(d-x)]}{Z_0\cosh(\Gamma d) + iZ\sinh(\Gamma d)} \tag{44}$$

From these equations the absorbed power density is obtained by applying Poynting's theorem [58] of energy conservation, which for fields with harmonic time dependence is:

$$-\frac{1}{2}j^*\cdot e = 2i\omega(u_m - u_e) + \nabla\cdot S \qquad (45)$$

u_e and u_m are the electric and magnetic energy densities and S is the complex Poynting vector

$$u_e = \frac{1}{4}e\cdot d^*, \quad u_m = \frac{1}{4}b\cdot h^* \qquad (46)$$

$$S = \frac{1}{2}(e\times h^*). \qquad (47)$$

The real part of the left-hand-side in Eq. (45) is the time average of the work done by the fields in the studied volume (Joule heat). This work is divided into the energy flow through the boundaries (Re($\nabla\cdot S$)) and the energy dissipated in the volume:

$$P = -2\omega Im(u_m - u_e) = -\frac{1}{2}\omega Im(\mu_0 h^*\cdot\tilde{\mu}\cdot h - \varepsilon_0 e\cdot\tilde{\mu}^*\cdot e^*). \qquad (48)$$

As $\tilde{\varepsilon} = \tilde{1}$ is real, the result for the power density in the sample is:

$$P(x) = \frac{1}{2}\omega\mu_0\mu_r'|h_y|^2 = 2\omega\mu_0\mu_r'h_0^2[\cosh 2\alpha(d-x) + \cos 2\beta(d-x)]\Big/$$
$$\left[1 - \frac{\alpha^2+\beta^2}{\varepsilon_r^2 k_0^2}\cos 2\beta d + (1 + \frac{\alpha^2+\beta^2}{\varepsilon_r^2 k_0^2})\cosh 2\alpha d + \frac{2\alpha}{\varepsilon_r k_0}\sin 2\beta d\right.$$
$$\left. + \frac{2\beta}{\varepsilon_r^2 k_0^2}\sinh 2\alpha d\right]. \qquad (49)$$

where $\quad k_0 = \omega\sqrt{\varepsilon_0\mu_0}$

The total power absorbed per unit area can be obtained by integration over the x-direction from $x = 0$ to $x = d$. The information accessible to the conventional FMR-experiment is the ratio of absorbed to incident power P_{abs}/P_{inc}:

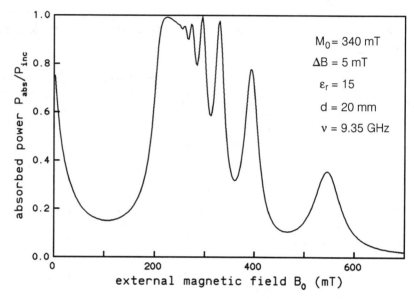

Fig. 14 Body-resonances in a 20 mm thick ferrite

Fig. 15 Body-resonances in a 2 mm thick ferrite

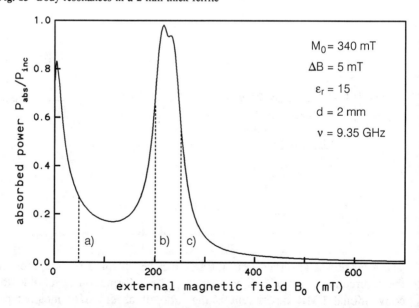

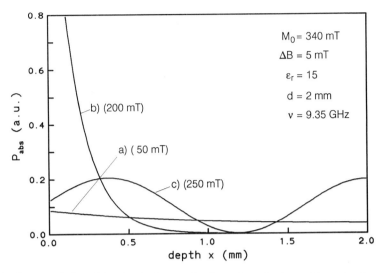

Fig. 16 Depth variation of the absorbed power density at several values of the external magnetic field.

Fig. 17 Depth dependence of the center of gravity of the absorbed power density distribution.

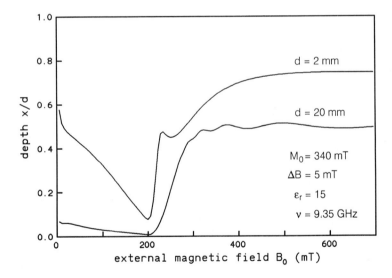

$$\frac{P_{abs}}{P_{inc}} = \int_0^d P(x)dx = \left(\frac{4\alpha}{\varepsilon_r k_0}\sin 2\beta d + \frac{4\beta}{\varepsilon_r k_0}\sinh 2\alpha d\right) \Big/$$
$$\left[\left(1 - \frac{a^2+\beta^2}{\varepsilon_r^2 k_0^2}\right)\cos 2\beta d + \left(1 + \frac{a^2+\beta^2}{\varepsilon_r^2 k_0^2}\right)\cosh 2\alpha d \right.$$
$$\left. + \frac{2\alpha}{\varepsilon_r k_0}\sin 2\beta d + \frac{2\beta}{\varepsilon_r^2 k_0^2}\sinh 2\alpha d\right] \tag{50}$$

Figures 14 and 15 show typical absorption spectra for thin ferrite plates of 20 mm and 2 mm, respectively. Besides the resonance peak at the internal resonance field $B_{res} = \omega/\gamma$ additional absorption characteristics appear. These extra peaks are frequently called "body-resonances" and are associated with multiple internal reflections which lead to standing waves within the sample. The number of body-resonances increases as the sample is made thicker or as the resonance linewidth of the ferrite is made smaller as long as the skin depth is sufficiently large. The condition for body-resonance is given by the well-known formula for the occurrence of optical interference

$$2d = (2n+1)\frac{\lambda}{2}, \tag{51}$$

where n is any positive integer, including zero, and λ is the wavelength within the ferrite which is connected to the phase constant β by $\beta = 2\pi/\lambda$.

In Figure 16 the power density distribution is plotted as a function of the x-coordinate for different values of the external magnetic field. Three different kinds of curves can be observed here. In the first case (trace a) the power absorption is nearly constant over the whole thickness of the sample corresponding to a volume heating. The second case represented in trace b where an exponential decay occurs resembles a surface heating. The third possibility (trace c) is an oscillatory behavior of the power density. The magnitudes of α and β determine which of the three cases occurs. Either the *cos* or the *cosh* term in Eq. (49) dominates. The *cosh* term is responsible for both the volume and the surface absorption case.

In Figure 17 the center of absorption is displayed for the two ferrite slabs. In the range of 0.2 T the changes are strongest. This gives rise to the supposition that around this field value there will be marked changes in the photothermal signal phase as shown further below.

3.2 *Photothermal FMR-signal of Thin Ferrite Slabs*

In this section we will derive the formulas for the photoacoustic (PA) and the photothermal beam deflection (PD) signal. As the absorption of microwave power inside the sample is spatially inhomogeneous the model for heat conduction has to take this into account [59].

In a fashion similar to the section before, we will deal with a one-dimensional model. To be able to include an arbitrary distribution this model consists of four layers (see Fig. 18). The layer g extending in the negative half space $x < 0$ is the gas above the sample. The space of the solid itself ($0 < x < d$) is divided into two sublayers denoted by s_1 and s_2 with same thermal properties. The boundary between s_1 and s_2 at x' comprises a δ-shaped heat source. The backing b is assumed to be infinite ($x > d$). The thermal parameters of the four regions are distinguished by their indices (g, s, b). So far the model will reproduce the temperature variation due to a unit heat source somewhere inside the sample. Switching to Greens' functions and integrating over the absorption distribution Eq. (49) this model will give the photothermal signal of a ferrite slab in front of a metal wall.

The equation of heat conduction that has to be solved in every region is given by

$$\frac{\partial \theta_i}{\partial t} - \alpha_i \frac{\partial^2 \theta_i}{\partial x^2} = 0 \quad \text{with} \quad i = g, s_1, s_2, b \ . \quad (52)$$

The ansatz for θ_i are thermal waves travelling from every boundary into the positive and negative x-direction. Here the time dependence has been omitted.

$$\theta_g = \Theta_1 \exp(\sigma_g x) \quad (x \leq 0)$$

$$\theta_{s1} = \Theta_2 \exp(-\sigma_s x) + \Theta_3 \exp[\sigma_s(x - x')] \quad (0 < x \leq x')$$

$$\theta_{s2} = \Theta_4 \exp[-\sigma_s(x - x')] + \Theta_5 \exp[\sigma_s(x - d)] \quad (x' < x \leq d) \quad (53)$$

$$\theta_b = \Theta_6 \exp[-\sigma_b(x - d)] \quad (x > d)$$

$$\text{with} \quad \sigma_{g,s,b} = (1 + i)\sqrt{\omega_m / 2\alpha_{g,s,b}} \ .$$

For θ_g and θ_b the boundary conditions of a finite temperature oscillation at infinity have already been taken into consideration. Further boundary conditions are the continuity of the temperature and the continuity of the heat flow at $x = 0, x', d$. At $x = 0, d$ for the heat flow this means

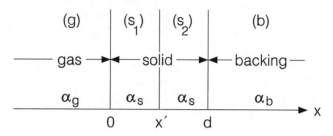

Fig. 18 The 1-D Model for heat conduction

$$\lambda_{g,s} \frac{\partial \theta_{g,s_2}}{\partial x} = \lambda_{s,b} \frac{\partial \theta_{s_1,b}}{\partial x} . \qquad (54)$$

At $x = x'$ an additional heat source f_0 has to be taken into account:

$$\lambda_s \frac{\partial \theta_{s_1}}{\partial x} = \lambda_s \frac{\partial \theta_{s_2}}{\partial x} + f_0 . \qquad (55)$$

Solving the linear homogeneous system of six equations we derive the temperature oscillations at the front and back surfaces Θ_1 and Θ_6

$$\Theta_1 = \frac{f_0}{\lambda_s \sigma_s}$$
$$\times \frac{(1-b)\exp[\sigma_s(x'-d)] + (1+b)\exp[-\sigma_s(x'-d)]}{(1+b)(1+g)\exp(\sigma_s d) - (1-b)(1-g)\exp(-\sigma_s d)} \qquad (56)$$

$$\Theta_6 = \frac{f_0}{\lambda_s \sigma_s}$$
$$\times \frac{(1+g)\exp(\sigma_s x') + (1-g)\exp(-\sigma_s x')}{(1+b)(1+g)\exp(\sigma_s d) - (1-b)(1-g)\exp(-\sigma_s d)} , \qquad (57)$$

where $g = \sqrt{\lambda_g \rho_g c_g}/\sqrt{\lambda_s \rho_s c_s}$, $b = \sqrt{\lambda_b \rho_b c_b}/\sqrt{\lambda_s \rho_s c_s}$. The Green's functions for this problem are connected to Θ_1 and Θ_6 by the following relations

$$G(x \leq 0, x') = \frac{\Theta_1}{f_0} \exp(\sigma_g x), \qquad (58)$$

$$G(x \geq d, x') = \frac{\Theta_6}{f_0} \exp[-\sigma_b(x-d)]. \qquad (59)$$

Temperature oscillations due to an absorption distribution $P(x')$ can now be calculated by integration over the product of P and a Green's function:

$$\theta(x) = \int_0^d P(x') G(x, x') dx' \qquad (60)$$

With the above Green's functions and assuming $P(x') = \frac{1}{2}\beta I_0 \exp(-\beta x')$ the well-known Rosencwaig-Gersho formula (9) in Ref. [30] can be reproduced.

For the present experiments the temperature oscillations at $x = 0$ and $x = d$ determine the signal. Using Eq. (49) and performing the integration in Eq. (60) the amplitude of the temperature oscillation at the front surface $x = 0$ yields:

$$\theta_g(0) = A \cdot \left\{ \frac{(1-b)\exp(-\sigma_s d)}{4\beta^2 + \sigma_s^2} [\sigma_s \exp(\sigma_s d) + 2\beta \sin 2\beta d - \sigma_s \cos 2\beta d] \right.$$

$$+ \frac{(1+b)\exp(\sigma_s d)}{4\beta^2 + \sigma_s^2} [-\sigma_s \exp(-\sigma_s d) + 2\beta \sin 2\beta d + \sigma_s \cos 2\beta d]$$

$$+ \frac{(1-b)\exp(-\sigma_s d)}{4\alpha^2 - \sigma_s^2} [-\sigma_s \exp(\sigma_s d) + 2\alpha \sinh 2\alpha d + \sigma_s \cosh 2\alpha d]$$

$$\left. + \frac{(1+b)\exp(\sigma_s d)}{4\alpha^2 - \sigma_s^2} [\sigma_s \exp(-\sigma_s d) + 2\alpha \sinh 2\alpha d - \sigma_s \cosh 2\alpha d] \right\} \qquad (61a)$$

with

$$A = (2\omega\mu_0\mu_r''h_0^2/\lambda_s\sigma_s)$$
$$\times[(1+b)(1+g)\exp(\sigma_s d)-(1-b)(1-g)\exp(-\sigma_s d)]^{-1}$$
$$\times\left[\left(1-\frac{\alpha^2+\beta^2}{\varepsilon_r^2 k_0^2}\right)\cos 2\beta d+\left(1+\frac{\alpha^2+\beta^2}{\varepsilon_r^2 k_0^2}\right)\cosh 2\alpha d+\left(\frac{2\alpha}{\varepsilon_r k_0}\right)\sin 2\beta d\right.$$
$$\left.+\frac{2\beta}{\varepsilon_r^2 k_0^2}\sinh 2\alpha d\right]^{-1} \quad (61b)$$

The temperature oscillation at the rear surface is given by:

$$\theta_b(d)=A\left\{\frac{(1+g)}{4\beta^2+\sigma_s^2}[\sigma_s\exp(\sigma_s d)+2\beta\sin 2\beta d-\sigma_s\cos 2\beta d]\right.$$

$$+\frac{(1-g)}{4\beta^2+\sigma_s^2}[-\sigma_s\exp(-\sigma_s d)+2\beta\sin 2\beta d+\sigma_s\cos 2\beta d]$$

$$+\frac{(1+g)}{4\alpha^2-\sigma_s^2}[-\sigma_s\exp(\sigma_s d)+2\alpha\sinh 2\alpha d+\sigma_s\cosh 2\alpha d]$$

$$\left.+\frac{(1-g)}{4\alpha^2-\sigma_s^2}[\sigma_s\exp(-\sigma_s d)+2\alpha\sinh 2\alpha d-\sigma_s\cosh 2\alpha d]\right\} \quad (62)$$

A is the same as in (61a).

The photoacoustic signal measured in a PA-cell is proportional to $\theta_g(0)$ and $\theta_b(d)$. In Figure 19 typical temperature oscillations at the front and at the rear side of the sample are displayed for the absorption distribution, Eq. (49).

The photothermal laser beam deflection signal normal to the sample surface is given by:

$$\Phi_n = \frac{1}{n_0}\frac{\partial n}{\partial T}\int_0^l \frac{\partial\theta}{\partial x}dy \quad (63)$$

The result for the ferrite in front of a short in the waveguide corresponds to the one for the PA-signal, with the inclusion of some multiplicative and phase factors. So the shapes of the PA- and PD-FMR signals are comparable.

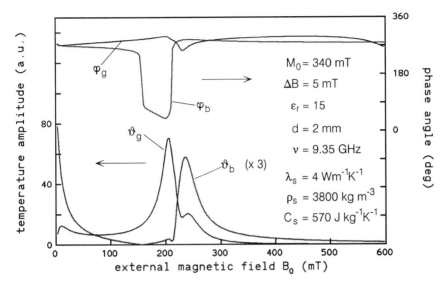

Fig. 19 Amplitude and phase of the temperature oscillations at the front ($\theta_g(0)$) and at the rear side ($\theta_b(d)$) of the sample.

3.3 Experimental Set-up

The experimental set-up used consists of a conventional EPR spectrometer which has been modified for PA- and PD-FMR spectroscopy. A block diagram of the set-up is shown in Fig. 5. In addition to the general description given in Sect. II.3 some details which are relevant for experiments discussed in this section are also given here.

The EPR spectrometer. The VARIAN E-line EPR spectrometer equipped with a microwave bridge working with a klystron in the X-band (8.8 -9.6 GHz) provides a maximum output power of 200 mW.

The 12" electro-magnet can produce a magnetic field up to 2.4T. The pole pieces used for the present investigations only allow a maximum value of 1.56 T. The magnetic field controller is interfaced to a personal computer.

Conventional FMR spectra can be recorded either using a field modulation technique or by directly measuring the incident and reflected microwave power. For the field modulation technique a cavity is used and the periodic signal is measured with a crystal detector in the microwave bridge and then fed to the spectrometer fixed-frequency lock-in amplifier. The direct measurement of the modulated signal is normally used along with a nonresonant waveguide arrangement and tunnel diodes. The lock-in and/or the diode signals are recorded by the PC with a 12 bit AD converter insert.

The PA spectrometer (Fig. 5). The extensions for the PA spectrometer are a modulator for the microwaves, a frequency variable lock-in amplifier and a waveguide photoacoustic cell.

The self-built PIN-diode modulator effects a square wave modulation with a 50% duty cycle. The modulation frequency must be supplied by an external oscillator. The bandwidth of the modulator is 20 MHz.

The lock-in amplifier is a fully computer controllable device (ITHACO-NF 3961) with an additional internal oscillator. The amplifier is controlled via an IEEE-bus. The data are recorded by the PC AD converter.

The PA waveguide cell (displayed in Fig. 6) consists of 0.5 mm thick cavity which has been routed out of a brass plate forming the end of a waveguide. From the cell a channel leads to a microphone chamber with a Knowles BT 1759 electret microphone.

The PD spectrometer (Fig. 5). The amplitude modulation of the microwaves is effected by the same modulator described above. In order to achieve higher excitation field strengths a HUGHES travelling wave tube amplifier (TWTA) is used behind the modulator. With this amplifier a maximum microwave output power of the spectrometer of approximately 20 W is obtained.

For the beam deflection technique a probe laser beam has to pass parallel to the sample surface at a small distance. This requirement can be fulfilled by a slotted waveguide: In the short sides of the waveguide walls a 1.3 mm high and 10 mm broad part is removed. This modification does not change the field distribution in the waveguide substantially. Using a double slit configuration with a slit above and below the sample made it possible to measure in "thermal reflection" and "thermal transmission".

The probe laser is a 1 mW HeNe-laser (632.8 nm). The choice of the laser is a vital point for the whole PD set-up. The UNIPHASE 1101P laser used in our laboratory fulfils the requirements as far as pointing stability, lack of periodic intensity modulations and low intensity fluctuations are concerned.

For the detection of the deflected laser beam a quadrant photodiode (CENTRONIC QD50-5T) is used. With this type of diode normal and transverse deflection can be recorded. For the different measurements two neighboring segments are coupled together and a preamplifier outputs the difference and sum voltages. In order to avoid low frequency disturbances a 4th order Bessel high pass filter can be used.

The probe laser and the detector are fixed on an optical bench below the air gap of the magnet. A 16 cm lens focuses the laser in the range of the sample. Spatially resolved measurements are possible in horizontal and vertical directions. For this purpose the waveguide arrangement is attached to a computer-controlled two-axis translation stage with a resolution of 2.5

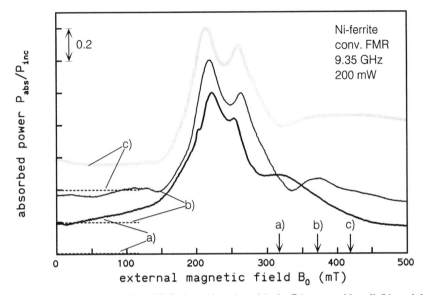

Fig. 20 Absorption spectra for a Ni-ferrite with a short (a), the PA waveguide cell (b), and the slotted waveguide along with the PA waveguide cell (c) behind the sample.

microns. Oscillations of the set-up are damped out very effectively within a few seconds.

Samples. Different kinds of ferrites have been studied: Ni-, NiZn- and LiZn-ferrites. The measurements presented here have been obtained with a 1.15 mm thick Ni-ferrite. The sample has the same cross-section as the X-band waveguide used: 23 x 10 mm^2.

3.4 FMR Measurements

The normalized spectra shown in Fig. 20 have been obtained at 9.35 GHz and 200 mW microwave power. The raw data are masked by a strong interaction of the sample with the microwave circuit when going through the resonance lines although a nonresonant waveguide arrangement was used. This is due to the sample size which is enormously large compared to the sample sizes normally used in FMR investigations. Besides the variation in the incident microwave power the amount of power lost in the microwave circuit between the microwave diodes and the sample had to be taken into consideration. So a normalization procedure based on the data on incident and reflected power had to be used.

In Fig. 20 conventional normalized absorption spectra for three configurations are shown. One spectrum has been obtained with a short behind the sample (trace a). The second trace in the figure is the result for the same sample but instead of using a brass plate directly behind the sample the PA-cell has been attached to the rear side of the ferrite. The third experiment has been performed with the same set-up as the second one and by adding the slotted waveguide. Strictly speaking, set-ups two and three do not correspond to the theory presented above, however, the real problem with a small layer of air between the sample and short is rather difficult to treat. Epstein [60] already pointed out that in this case higher order modes occur behind the sample. The experimental results confirm that in a first approximation the theory applied here works satisfactorily.

Comparing the three spectra in Fig. 20 one can state that the agreement for the low order body-resonances is good, while for the higher order resonances the absorption for the PA-cell case is lower than for the brass plate case and slightly shifted to higher field values. Adding the slotted waveguide the line-shift becomes larger and the absorbed power decreases. This result is confirmed by measurements with other ferrites. An explanation of this behavior can be given by the additional layer of air between the ferrite and the short, where a certain amount of the electromagnetic energy can leave the ferrite. Therefore, at a fixed value of the external magnetic field part of the energy that would be absorbed in the case considered in the theory is not available to be absorbed.

In conclusion, the conventional measurements show the expected

Fig. 21 PA-FMR amplitude and phase of a Ni-ferrite for a microwave frequency of 9.35 GHz and a modulation frequency of 107.7 Hz.

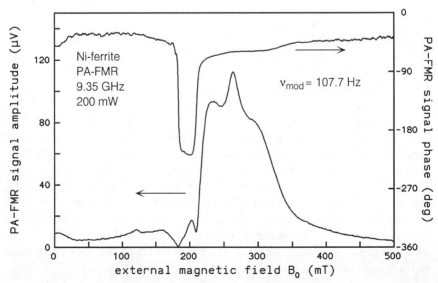

occurrence of "body-resonances". Although the experimental set-up does not match fully the theoretical conditions there exists a qualitative agreement between experimental and theoretical results.

3.5 PA-FMR Measurements

The spectra have been recorded at a microwave frequency of 9.35 GHz and at different modulation frequencies. Here, too, a normalization had to be performed as the incident microwave power was not constant during the field scan [61]. The normalization procedure was as easy as that used in the former section because the PA-FMR signal varies proportional to the incident microwave power.

The PA-FMR signal trace (Fig. 21) differs markedly from the conventional FMR spectra in Fig. 20. The main peak here occurs at 0.26 T, whereas at about 0.2 T, where the conventional signal has its maximum, the PA signal amplitude is very small. The most striking feature of the PA spectra is the subsidiary peak a 0.2 T. At this value of the external magnetic field there is also the biggest change in the phase signal which is a measure for the runtime of the thermal waves. Depending on the modulation frequency phase lags between 155° and 175° can be observed indicating that the signal at 0.2 T is produced near the front side of the ferrite which is exposed to the microwaves.

A comparison of the experimental results with the theoretical data

Fig. 22 PD-FMR amplitude and phase of a Ni-ferrite measured in thermal reflection in the middle of the sample (a), and 100 μm away from the sample edge (b).

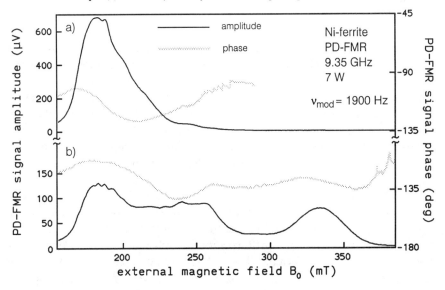

obtained from Eq. (62) yields a good qualitative agreement. The two peaks in the amplitude signal and the phase lag can be reproduced although the feature at high frequencies cannot be described by the theory. One possible origin for the quantitative disagreement has already been pointed out in the last section. In addition, the sample thickness is not negligible compared to the lateral dimensions. For that reason there could be an anisotropy in the heat flow, which is not included in the one-dimensional heat conduction problem.

PA-FMR measurements at different modulation frequencies reveal the influence of the inhomogeneous demagnetizing field inside the sample. For higher modulation frequencies the peaks of the absorption spectra shift to higher values of the external magnetic field because the demagnetizing field, directed opposite to the external field, causes a rise in the inner resonance field. This increase is larger for points near the sample surface. For larger modulation frequencies, i.e. for smaller thermal diffusion lengths, only a thinner layer next to the sample surface is seen by the PA-detection.

Frequency dependent measurements at the main peak at 0.26 T have been performed in the range from 6 Hz to 6 kHz. The maximum signal amplitude can be observed between 20 and 30 Hz. In a log-log plot of signal versus modulation frequency the trace does not have a constant slope due to the transmission characteristic of the microphone. Beyond 100 Hz the amplitude decreases proportional to $\omega^{-1.1}$. This relationship is what can be expected from theoretical considerations for the PA-signal which result in a dependence between ω^{-1} and $\omega^{-3/2}$. At approximately 3 kHz the slope of the curve decreases and at 5 kHz Helmholtz cell resonances occur.

3.6 PD-FMR Measurements

The PD-FMR measurements presented in this section have been recorded at 9.35 GHz microwave frequency and modulation frequencies between 6 Hz and 40 kHz. The probe beam radius above the sample was 110 μm and the distance between the sample and the probe beam center was 85 μm. The typical spectra shown here have been obtained with the probe laser beam centered in the middle of the sample. For spatially resolved measurements the laser beam position was varied in steps of 150 μm. PD-FMR spectra have been recorded in thermal reflection and in thermal transmission.

The spectra in Fig. 22 have been measured with the laser beam above the sample, i.e. in thermal reflection, a measurement not accessible to PA-FMR. The signal from the middle of the sample shows a peak at 0.19 T. In this field region in the PA-FMR spectra the subsidiary peak in connection with the phase lag occurred. The PD signal observed here supports the finding of the former section that the 0.2 T peak results from power absorption near the front surface of the ferrite. The PD signal phase indicates

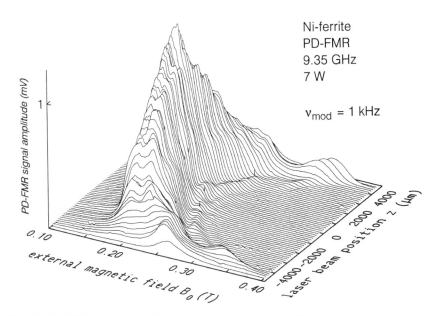

Fig. 23 PD-FMR spectra of a Ni-ferrite measured in thermal reflection.

Fig. 24 PD-FMR signal measured in thermal transmission in the middle of the sample (a) and 2 mm away from the margins (b).

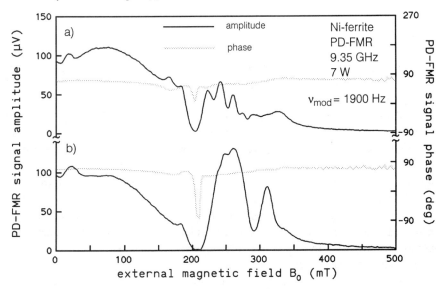

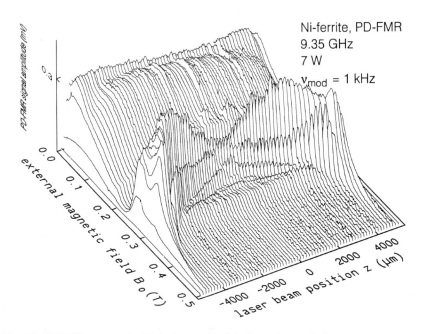

Fig. 25 PD-FMR spectra of a Ni-ferrite measured in thermal transmission.

that for higher fields the signal arises from deeper inside the sample as can also be seen from Fig. 17. The second signal exhibits several body-resonances under the influence of the inhomogeneous demagnetizing field. The influence of the demagnetizing field for PD-FMR measurements is much more important that for PA-FMR studies, as the PD-FMR measurement only averages the signal in that region of the sample which lies under the probe laser beam. For that reason in PA-FMR measurements only the effect of the demagnetizing field in the x-direction can be investigated, whereas in the PD-FMR measurements the laser beam position allows the discrimination of the effect of the demagnetizing field in the z-direction. Comparison of the observed line shifts with a theoretical calculation of the inhomogeneous demagnetizing field in a slab [62] reproduces the correct results.

This is clearly demonstrated in Fig. 23 where spectra for laser beam positions 150 µm apart have been recorded at a modulation frequency of

1 kHz. The decrease of the maximum amplitude from the middle of the sample in the direction towards the margin cannot be understood, as measurements of the high frequency field strength yield a constant value over the whole width of the waveguide.

A PD-FMR spectrum measured in thermal transmission in the middle of the sample is shown in Fig. 24. In this spectrum, as in the spectra measured at other laser beam positions, at 0.2 T, a minimum — but no subsidiary peak as in the PA-spectra — can be observed. Signals near the margins of the sample exhibit two peaks at 0.26 T and 0.31 T as measured in the conventional signal. Signals arising from the middle of the sample exhibit a structured peak the shape of which changes with changing modulation frequency. Demagnetization effects are visible as clearly as in the reflection measurements. The maximum signal amplitudes in the spectra measured in the middle of the sample are smaller than in spectra from the margins. This coincides with the fact that in reflection measurements the maximum amplitude is observed in the middle of the sample meaning, that a great amount of microwave power has already been absorbed near the front surface. Figure 25 shows the spatial variation of the PD-FMR spectra with the same resolution as in Fig. 23.

3.7 Conclusion

In the preceding section inhomogeneous absorption of microwave radiation was investigated by means of the PA- and the PD-FMR technique. The theoretical model used is based on one-dimensional propagation of electromagnetic waves in an air-ferrite system with a short behind the sample. Main features of this model are "body-resonances" which were observed with conventional FMR techniques. The thermal signals were derived using a one-dimensional model for the heat conduction applying a Green's function technique to take into account the inhomogeneous distribution of absorbed power. A comparison of experimental and theoretical photothermal data showed that principal features could be reproduced by the theory. A quantitative analysis was not possible because of the vast influence of the inhomogeneous demagnetization field in the sample. The demagnetizing field causes a shift of the resonance line positions, which were studied for different positions in the sample. The observed line shifts could be identified by comparison with the theoretical predictions for the demagnetizing field in a slab. The discrepancies between theory and experiment can be understood if one takes into account that, first, the geometry assumed for the theoretical considerations could not exactly be reproduced in the experiment and, second, both the one-dimensional electromagnetic and the heat diffusion problem are not sufficient for the description of the present anisotropic problem. But the experimental data given in this work show that especially the PD-FMR technique is capable of imaging spatially inhomogeneous absorption distributions in different depths of samples. This is a feature which is not accessible to conventional FMR measurements as the spectra presented in this chapter clearly demonstrate.

4. Magnetostatic Modes in Yttrium Iron Garnet (YIG)

Long-wavelength magnetostatic modes (MSM) propagating in a ferromagnetic sample of finite dimensions have been studied for many years. The first observation of such modes was reported by White and Solt [63] in spheres in 1956. The theoretical explanation was published one year later by Walker [64], after whom MSM in spheroids have been named. Damon and Eshbach (DE) [65] were the first who predicted MSM in ferromagnetic slabs in a tangential external magnetic field. The first experimental observation of those modes was reported in 1968 by Brundle and Freedman [66].

Here we restrict the detailed theoretical description to the modes predicted by DE because the majority of the experiments discussed below are performed on slabs. In addition, a great number of properties of the modes in slabs can be found again in the case of the spherical geometry. A further invaluable advantage of the slab geometry is its simple form of solutions which allow an intuitive understanding of the complex problem of magnetostatic modes.

For 30 years various aspects of the nature of MSM in spheres, slabs and films — predominantly in magnetic garnets — have been investigated because of interesting technological applications in microwave devices [67] and the physical processes accompanying their propagation in ferromagnetic materials [49,68]. Physical effects studied along with MSM propagation comprise their dispersion, influence of the anisotropy of the magnetic media, different boundary conditions and many other properties as a recent review [69] demonstrates. In recent years MSM gained renewed interest from the point of view of nonlinear dynamics [70,71].

First spatially resolved investigations of YIG samples have been carried out by Dillon [15] who succeeded in obtaining photos of MSM in ferromagnetic $CrBr_3$ films at low temperatures making use of the Faraday rotation. The first photothermal observation of MSM in YIG was reported by Wettling et al. in 1981 [72], who detected Walker modes in a PA-FMR experiment in order to demonstrate the applicability of the PA-FMR technique to MSM. In 1985, Vlannes [73] used micro-coils to probe the rf magnetization distribution in YIG films. Photothermal laser beam deflection experiments and photothermally modulated FMR on MSM in YIG were first conducted by von Geisau et al. in 1989 [74,75] and by Hoffmann et al. [47] in 1991, respectively. Rezende [16], applied Brillouin light scattering to record magnetostatic mode patterns in YIG films. For spheres, to the knowledge of the authors, the only attempts to visualize the rf field distribution are the ones in which the authors are involved [76,77].

In the following sections we will describe the magnetostatic theory for YIG slabs and spheres, as well as our investigations on both kinds of samples based on the papers cited above.

4.1 Magnetostatic Modes in Slabs

The DE-theory [65] describes the solution of a magnetostatic problem in a slab of thickness s, which is infinitely extended in the y- and z-directions (Fig. 26). The external magnetic field H_0 is applied tangential to the sample surface in the z-direction. The problem requires the simultaneous solution of Maxwell's equations $\nabla \cdot B = 0$ and $\nabla \times H = 0$ with their respective boundary conditions and of the equation of motion for the magnetization M neglecting losses. Because of the curl-equation the problem can be solved by a magnetic scalar potential ψ with

$$h = \nabla \psi \qquad (64)$$

where in rectangular coordinates the differential equation separates:

$$\psi = A\exp(\pm ik_x x)\exp(\pm ik_y y)\exp(\pm ik_z z) \qquad (65)$$

Fig. 26 Coordinate system for the DE-theory

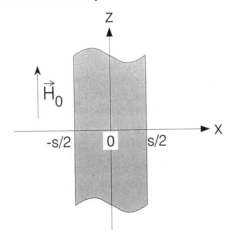

Inside the sample the potential ψ^i must satisfy

$$(1 + \kappa) \left(\frac{\partial^2 \psi^i}{\partial x^2} + \frac{\partial^2 \psi^i}{\partial y^2} \right) + \frac{\partial^2 \psi^i}{\partial z^2} = 0 \qquad (66)$$

and outside Laplace's equation

$$\nabla^2 \psi^e = 0 \qquad (67)$$

has to be fulfilled. The solution of the boundary value problem results in three equations in the five variables ω, k_x^i, k_x^e, k_y and k_z:

$$(1 + \kappa)(k_x^{i\,2} + k_y^2) + k_z^2 = 0 \qquad (68)$$

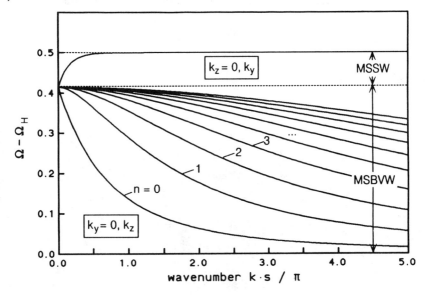

Fig. 27 Dispersion of magnetostatic modes in a transversely magnetized unbounded slab of thickness s for pure mode propagation. The volume modes propagate along the z-direction ($k_y = 0$) and the surface mode propagates along the y-direction ($k_z = 0$).

$$k_x^{e^2} - k_y^2 - k_z^2 = 0 \tag{69}$$

$$k_x^{e^2} + 2k_x^e k_x^{i}(1+\kappa)\cot(k_x^{i}s) - k_x^{i^2}(1+\kappa)^2 - (\nu k_y)^2 = 0 \tag{70}$$

where κ and ν are the susceptibility tensor components, here written as functions of the dimensionless frequencies $\Omega_H = B_i/M_0$ and $\Omega = \omega/\gamma M_0$.

$$\chi_{xx} \equiv \kappa = \frac{\Omega_H}{\Omega_H^2 - \Omega^2} ; \quad \chi_{xy} \equiv \nu = \frac{\Omega}{\Omega_H^2 - \Omega^2} \tag{71}$$

B_i is the effective internal magnetic field.

With the elimination of two variables the permitted solutions are lying on hyper-surfaces in a three-dimensional parameter space. The appropriate representation for MSM is the one in a ω, k_y, k_z-space. Its dispersion relation has the form

$$(1 + \eta^2) + 2|(1 + \eta^2)^{1/2}| \left(-\frac{1 + \eta^2 + \kappa}{1 + \kappa}\right)^{1/2}$$

$$\times (1 + \kappa)\cot\left[|k_y|s\left(-\frac{1 + \eta^2 + \kappa}{1 + \kappa}\right)^{1/2}\right] \tag{72}$$

$$+ (1 + \kappa)^2 \left(\frac{1 + \eta^2 + \kappa}{1 + \kappa}\right) - \nu^2 = 0$$

where $\eta = k_z / k_y$.

The dispersion relation, Eq. (72), has two physically different kinds of solutions. This is due to the cotangent term which may have either pure real or pure imaginary arguments.

In the first case, when $(1+\kappa) < 0$ there is an infinite number of solutions because of the periodicity of the cotangent. From Eq. (68) it becomes clear that k_x^i, the wavevector component inside the slab, is real and therefore this case corresponds to the propagation of volume modes, which are designated as magnetostatic backward volume waves (MSBVW), where "backward" refers to their negative group velocity. For $k_y = 0$ the relation Eq. (72) assumes a rather simple form in the variable $\alpha = [-1/(1 + \kappa)]^{1/2}$

$$2\cot[|k_z|s\alpha] = \alpha - \frac{1}{\alpha}, \tag{73}$$

which is displayed in Figure 27. Volume modes can be observed between the frequencies

$$\Omega_1 = \Omega_H \tag{74}$$

and

$$\Omega_2 = (\Omega_H^2 + \Omega_H)^{1/2}. \tag{75}$$

If, on the other hand, $(1+\kappa) > 0$, the argument of the cotangent is imaginary and the cotangent can be replaced by a hyperbolic cotangent. Therefore, there exists only one solution in this case. This kind of solution is a magnetostatic surface wave (MSSW) as k_x^i is imaginary, which leads to a decreasing potential inside the sample. Surface wave modes propagate between the upper frequency limit of volume waves Ω_2 and the maximum frequency

$$\Omega_3 = \Omega_H + 1/2. \tag{76}$$

Another important difference between MSSW and MSBVW is the fact that while MSBVW can propagate for any value of k_y and k_z the propagation of MSSW is only allowed if $|\eta| < \Omega_H^{-1/2}$. For $k_z = 0$ the relation (72) again assumes a very simple form

$$\Omega^2 = \Omega_H^2 + \Omega_H + [2 + 2\coth(|k_y|s)]^{-1} \tag{77}$$

which is also plotted in Figure 27.

Up to this point only laterally infinite samples have been treated. In a slab with finite lateral dimension b in y-direction and c in z-direction, the boundary conditions imposed on the system require that due to a geometrical quantization the wave numbers k_y and k_z assume discrete values $k_y = \pi n_y/b$, $k_z = \pi n_z/c$, resulting in standing wave modes denoted by the indices (n_y, n_z). Now the description of MSM will depend on the dimensions of the slab.

If the lateral dimensions are such that the values of k_y and k_z are not much larger than the wavenumbers of electromagnetic waves propagating in a ferromagnet the magnetostatic approximation is no longer valid. In this case, k_y, k_z being of the order of $k_0 = \omega/c\sqrt{\varepsilon_r}$, where ω is the frequency of the wave, ε_r is the relative dielectric constant, and c is the velocity of light, for the description of MSM in the ferromagnetic slab the full Maxwell's equations (including retardation terms) should be used. Karsono and Tilley [78] and Ruppin [79] considered theoretically the dispersion characteristics of such magnetodynamic waves. They have found that the main discrepancy between magnetodynamic modes and DE waves exists in the limit of long wavelengths where k_y, k_z are of the order of k_0 and that this discrepancy grows for thick slabs.

Further deviations from the DE behavior can arise from magnetocrystalline anisotropy [80]. For instance, for YIG ‖ [100] the magnetic anisotropy field B_{an} is of the order -8 mT. This can lead to strong changes in the spectra of MSM. In particular, it was found in Refs. [81,82] that the top and bottom boundary frequencies for the MSM spectra Ω_1 and Ω_3 are shifted at essential values. The top frequency for MSBVW, Ω_2, which coincides with the bottom frequency for MSSW in the case of an isotropic ferromagnet, splits up to two frequencies and a band of anisotropic volume modes appears. These modes under certain conditions can be backward or forward modes with opposite group velocities. They were found experimentally by Zavislyak [82].

All previous statements concerned MSM propagation at relatively small microwave power when the small signal approximation is still valid and the influence of excited modes on others can be neglected. But if the power of the excitation grows and exceeds a certain threshold value, nonlinear multimagnon processes may occur. Suhl [83] has classified them into first-order (three-magnon) and second-order (four-magnon) processes.

The first-order processes occur above a relatively small threshold power because only three waves take part in the process. Very often such three-magnon processes restrict the power of the excited waves and the four-magnon processes cannot be reached. If the three-magnon processes are forbidden by conservation laws, four-magnon processes can set in at high microwave power. Its threshold power increases quadratically with the thickness of the ferromagnetic layer [84]. This means that for thick YIG slabs MSMs can be considered as linear even if the power is not very small. These conditions make linear and nonlinear regimes of MSM in YIG slabs accessible to locally resolved investigations even using PD-FMR.

The theory of instability of intensive monochromatic spin waves in an infinite ferromagnet was considered by Zakharov *et al.* [85]. Experimentally, three- and four-magnon decay processes for MSSW in thin ferromagnetic films were investigated in [86,87] and forward volume MSW in Refs. [88,89]. Theoretically, four-magnon decay processes have been studied for volume MSW in Ref. [89] and for MSSW propagating in thin ferromagnetic films in Ref. [84].

4.2 *Experimental Set-up for Slab-shaped Samples*

Experiments on YIG samples have been performed with the conventional FMR technique as well as with the PD- and PM-FMR detection schemes.

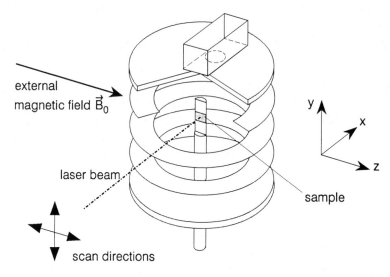

Fig. 28 The light-access cylindrical TE_{011} cavity (Q = 2100).

The PD spectrometer. The PD spectrometer has already been described in Sections II.3 and III.3.3 For the investigation of YIG it only differs in some points. The sample does not have the waveguide dimensions and measurements are only possible at the top surface, i.e. the surface pointing towards the incoming microwaves.

The PM spectrometer. The PM spectrometer is used in connection with the conventional EPR spectrometer as described in Sect. III.3.3.

For PM-FMR measurements a light-access cavity has to be used. For this purpose an open cylindrical TE_{011}-cavity with an unloaded Q of 2100 has been built (see Fig. 28). When this cavity is equipped with additional Helmholtz modulation coils both conventional and photothermal experiments can be conducted with the same set-up.

The local temperature modulation of the sample is effected by a SIEMENS LGK 7653 HeNe laser (632.8 nm, 12 mW), which is square wave-modulated with an acousto-optic modulator ISOMET IMD 80. The bandwidth of this system at the moment is restricted by the narrowband detection system to 120 kHz. But in the near future this will be extended to 10 MHz. The laser is focused onto the sample with a 17 mm lens. The minimum focal spot diameter obtained is 20 µm.

Spatial resolution is achieved by moving the cavity with respect to the optical set-up with a step width down to 2.5 μm.

The photothermal signal is measured in the conventional signal path of the spectrometer. After the microwave diode it is preamplified and the signal is analyzed by an external vector-lock-in amplifier.

Samples. Various kinds of YIG samples have been studied: Bulk samples of 1 mm thickness and 8 mm by 4 mm lateral dimensions, films of 17 μm thickness and the same lateral extent and films of 15 μm thickness with 675 μm by 590 μm lateral dimensions. All samples were single crystals. The bulk samples were grown from a flux [90] and the films were grown by liquid phase epitaxy [92]. The samples treated in this chapter have been supplied by S.M. Rezende (Recife, Brazil; bulk sample), W. Tolksdorf (Philips Hamburg, Germany; films) and by P. de Gasperis (Rome, Italy; films).

Fig. 29 Conventional FMR spectrum of the YIG bulk sample.

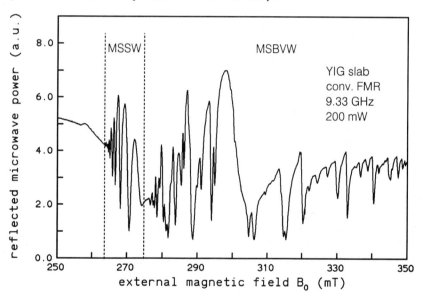

4.3 PD-FMR Study of a YIG Bulk Sample

Measurements at low microwave power. The experiments reported in this section were performed on a YIG single crystal of dimensions $8.4 \times 4.4 \times 0.98$ mm^3 [74]. Its large face is (110), the long side of the crystal is cut along [3$\bar{3}$2]. A conventional FMR spectrum recorded at a microwave power of 200 mW is shown in Fig. 29.

For several reasons it is rather difficult to deduce from the conventional FMR spectrum the mode numbers for the MSM. First, for the sample size investigated here the pure magnetostatic approximation is not valid. So the full Maxwell's equations have to be taken into account [79]. Second, also because of the large sample size the demagnetizing field inside the sample can no more be assumed to be homogeneous. The inhomogeneous demagnetizing field [62] affects the propagation properties of MSM in the slab and now makes the dispersion relation nonuniform in space.

The consequences which result from this can be seen in the two figures which show the MSSW (Fig. 30) and the MSBVW part (Fig. 31) of the spectrum, respectively, recorded with the laser beam deflection technique at a microwave power of 200 mW and a modulation frequency of 1.5 kHz corresponding to a thermal diffusion length of 65 μm in air and 110 μm in YIG. The maximum signal amplitude obtained in these measurements was 100 μV.

The distinction between surface and volume wave modes is quite easy. As the most intensive MSSW are those with mode number $(n_y,1)$ one expects no pronounced variation in the z-direction, but several modes occupying a range of 11 mT. Therefore, the identification of these modes with the laser pointing along the z-direction is not possible. For MSBVW the most intense modes are the $(1,n_z)$ ones which can be easily identified by counting the number of maxima of the PD signal in the z-direction.

The PD-signal in the range of the MSSW exhibits a growing phase lag for increasing external field B_0 for all except the three modes with the lowest n_y [93]. This means that the modes at the low field limit cling more closely to the sample surface than those at the other end of the MSSW mode spectrum: It takes a longer time for the thermal waves to reach the sample surface when the layer contributing to the signal is thicker. This result is justified by the DE theory which predicts higher wavenumbers for the low field modes.

One of the effects of the inhomogeneous demagnetizing field mentioned above can be observed in the MSSW spectrum. As can clearly be seen in Fig. 30, the surface wave modes do not extend over the whole of the sample but are confined to the center portion of the slab. This can be understood when analyzing the inhomogeneous demagnetizing field which is rather constant in the middle of the sample, but in the y-, as well as in the z-direction, sharply decreases and increases, respectively, when approaching the

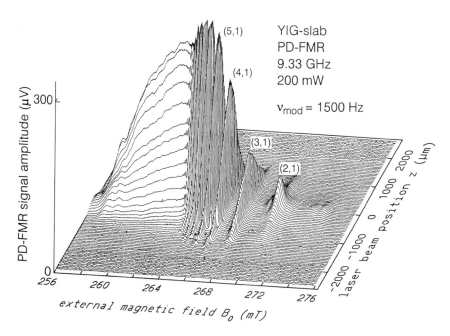

Fig. 30 PD-FMR spectra of MSSW in the YIG bulk sample.

Fig. 31 PD-FMR spectra of MSBVW of the YIG Bulk Sample. Only the indices of the most strongly excited modes are given.

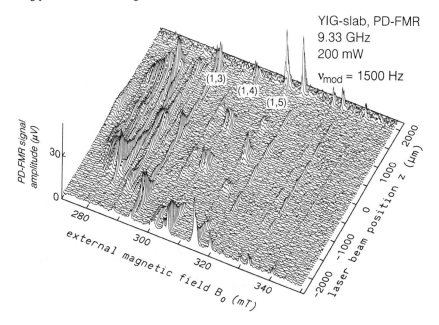

sample margins. Due to this field which is included in the internal field occurring in the dispersion relation, those parts of the sample which sense a strong field gradient may be exposed to a field value which does not lie within the limits of MSM propagation while the center part of the sample is in resonance [93]. For YIG films the error due to demagnetizing effects can be removed by assuming effective sample dimensions a few percent smaller than the real ones [94]. For a bulk sample this procedure is not possible because a virtual surface, at a fixed field, surrounds that part of the sample where resonance may occur and does not at all resemble a slab but is markedly curved.

The field gradient due to the inhomogeneous demagnetizing field has a second effect which can be seen in the MSBVW spectrum (Fig. 31), e.g. in the (1,5) mode. As the internal field even changes slightly in the region where resonance is still possible, this is equivalent to a non-constant wavevector along the z-direction. For that reason, the potential ψ^i cannot be given in a simple exponential form such as $\psi^i \propto \exp(-k_z z)$ which then allows to assume a geometrical quantization $k_z = n_z \pi / c$ for the wavenumber k_z. Using a WKB-like approximation the modified quantization condition yields $\int k_z(z) dz = n_z \pi$ [94]. The modified potential can explain the different spacings between the maxima in the (1,5) mode.

An effect which is due to crystalline anisotropy can be observed when the mode at 274 mT is considered. Because of the above-mentioned

Fig. 32 Frequency dependent normalized PD-FMR spectra of MSSW and an anisotropy mode in the YIG bulk sample obtained in the middle of the sample.

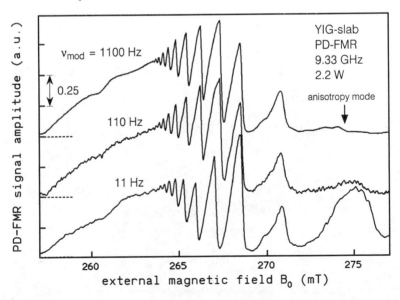

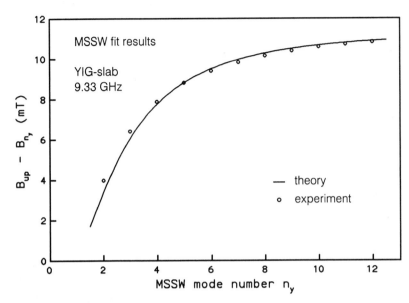

Fig. 33 Fit for the MSSW line positions observed in the YIG bulk sample. B_{up} is the upper field limit of the MSSW corresponding to Ω_2, Eq. (75).

Fig. 34 Fit for the MSBVW line positions observed in the YIG bulk sample. B_{low} is the lower field limit of the MSBVW corresponding to Ω_2, Eq. (75).

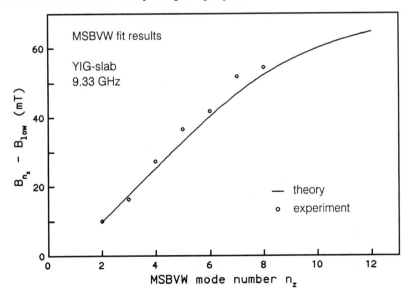

problems, from theory it is not clear from the beginning whether the mode in questions still belongs to the surface waves or is already a volume wave. The depth resolving capabilities without doubt are able to show that it is definitely a volume mode. Frequency dependent measurements (Fig. 32) showed a characteristic behavior of the investigated mode amplitude: For lower frequencies its relative amplitude compared to that of the next surface mode is higher than for higher frequencies, thus indicating that the main contribution of the signal is produced in the bulk.

The feature at 275 mT is attributed to the uniform mode which is excited in a field gradient. In addition, some of the modes, due to anisotropy, may also be of the backward wave type propagating along the z-direction [95], a topic which is still under investigation.

A fit for the line positions of MSSW and MSBVW including retardation effects and a variable reduction for the effective dimensions depending on the external field is given in Figs. 33 and 34. The agreement between theory and experiment for this difficult case is fairly good [93].

The experimental data on YIG shown up to this point were all performed with the laser beam probing variations in the z-direction. Due to technical reasons it is not possible to obtain good spectra for the other geometry, when the laser is scanned parallel to the y-direction, as for this experiment the long side of the waveguide walls has to be modified to allow the laser to pass over the sample. When this is done wall currents in the waveguide walls are cut which leads to a marked distortion of the rf field inside the cavity and a leakage of microwave power into the surroundings.

Fig. 35 PD-FMR spectrum of MSSW in the YIG bulk sample for microwave power of 2.2W (a) and 200 mW (b). The laser beam was positioned in the middle of the sample.

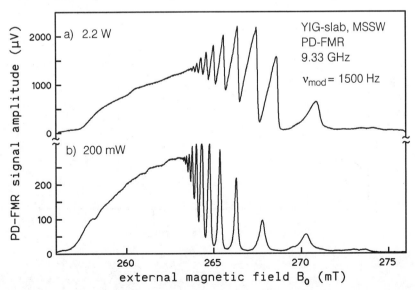

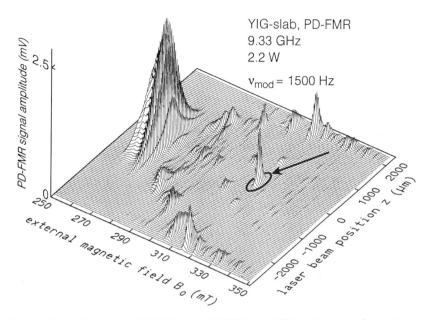

Fig. 36 PD-FMR spectra of MSSW and MSBVW in the YIG bulk sample for a microwave power of 2.2 W.

With this configuration some test spectra have been measured showing the right behavior of the variation of the PD-FMR signal. Nevertheless, with the signal-to-noise-ratio of the data recorded with this arrangement being very poor, these data are therefore omitted here [93]. An alternative experimental solution to this problem is offered by the PM-FMR which will be discussed in the next section.

Measurements at high microwave power. When the incident microwave power is increased above 200 mW there are marked differences of the thus obtained spectra from those for the lower microwave powers [75].

The first observation that can be made is the fact that a foldover-like effect in the MSSW occurs. This is clearly demonstrated in Fig. 35. The trailing edges of the surface modes exhibit a very steep descent, whereas the 200 mW measurement shows nearly symmetrical curves. This behavior is related to a shift of the resonance pattern to higher magnetic fields. From the form of the curves one could, at first thought, interpret these findings as a foldover-effect due to a bistability of the z-component of the rf magnetization as for example described by Lax (Ref. [49] Sect. 5.3). But when the resonance lines are passed through with different scan velocities the absolute line shift turns out to be a function of that quantity. The explanation for the line shift and the form of the trailing edge can be given in terms of the temperature dependence of the saturation magnetization M_0 of the YIG sample. At room temperature there exists a nearly linear dependence of the value of M_0 of the temperature and the temperature coefficient is approximately -0.36 mT/K. In the course of the photothermal experiment when a resonant line is reached with the absorption of microwave power and the associated temperature rises in the sample, the value of M_0 is decreased. From the form of the boundary frequencies Ω_2 Eq. (75) and Ω_3 Eq. (76) for MSSW it is easily seen that for lower M_0 the externally applied field has to grow to enable resonance. When the external field passes through resonance, the temperature in the sample rises until the external field value coincides with the peak position of the continuously shifted resonance line. After this point the amount of absorbed microwave power decreases as the line position does. With this shift the trailing edge of the foldover-like line falls to small values nearly immediately.

The second difference in the 200 mW spectra is the intensity of the (1,5) mode which does not exhibit any anomalous behavior for measurements below $\approx$ 1 W. Above that value the intensity of that mode grows dramatically whereas the other modes retain approximately the same intensity. The intensity gain of this particular mode points towards a strongly enhanced energy flow into this mode due to a four-magnon effect. The power threshold for this effect estimated after the calculations in Ref. [84] is consistent with the measured threshold value [93].

Conclusion. In the case of YIG slabs the PD-FMR technique has proven its capability of detecting spatially resolved MSM spectra. In comparison to the conventional spectra the additional information obtained on the character of the observed modes — not accessible to any other method — helped to identify individual modes. Apart from this, with the help of the PD-FMR a new physical effect due to four-magnon interaction could be observed: the flow of energy into the (1,5) mode at high microwave powers.

Besides these positive results, the high microwave power irradiated onto the sample and absorbed in the course of resonance absorption is a very critical matter when judging the capabilities of photothermal microwave techniques. With a high power absorption the thermal loading of the sample cannot be neglected even if foldover-like effects cannot be seen at 200 mW microwave power. In order to avoid influences of temperature effects the driving rf power should always be chosen as small as possible. The conventional signal could still be recorded at a microwave power of 5 mW whereas 100 mW is the absolute minimum rating for a PD-FMR experiment using a non-resonant waveguide arrangement.

In order to overcome this disadvantage, in the following part measurements of YIG films with conventional and PM-FMR are presented.

4.4. PM-FMR Study of YIG Film Samples

In this section measurements of two YIG film samples will be presented. Sample A is a 7.2×3.8 mm$^2 \times 17$ µm film. Sample B is a $675 \times 570 \times 15$ µm^3 film. Both samples are (111) films with the long side parallel to [1$\bar{1}$0].

The PM-FMR experiments at room temperature presented here have been conducted with the cylindrical cavity. Although the filling factor of the cavity with sample A is considerably reduced compared to the bulk sample treated before impedance matching of the loaded cavity to the microwave circuit — a sine-qua-non for resonant FMR experiments — could only be achieved for a limited range of power and magnetic field variation. For this reason, the cavity was slightly mistuned so that the distortion of the rf field inside the cavity had no negative effect on the mode pattern. With sample B again being markedly smaller than sample A perfect cavity matching could be obtained. Nevertheless, even with sample B, experiments could only be performed at low microwave powers. In this case low power means an incident radiation of approx. 0.4 to 0.5 mW corresponding to a maximum induction in the waveguide of 0.14 to 0.16 µT [37] which has to be multiplied by the quality factor to obtain the induction at the location of the sample.

Additional experiments at variable temperatures have been undertaken to get new information on the signal generation process in PM-FMR. For these measurements a rectangular TE_{103} light-access cavity with dewar insert has been been used. With this equipment PM-FMR and conventional FMR experiments from 4 to 600 K have been performed.

Imaging of YIG films. A set of PM-FMR spectra is displayed in Figs. 37 and 38 for a scan where the y-position of the laser was varied and the z-direction was fixed at $z = 0$ in the middle of the sample [47]. The mode density for a film with the same lateral dimensions as a 50 times thicker bulk sample is larger than expected from the DE theory.

With the PM-FMR technique more than 50 surface wave modes could be observed, 35 of which could only be used for comparing the obtained results with the DE theory as the intensity was too low for other reliable results for n_y. The mode number n_y can be obtained by using a Fourier transform which returns the spatial frequency (i.e. the mode number) at a fixed field B_0. The respective scan in the z-direction confirms that $n_z = 1$ as there is always only one peak of the PM-FMR signal.

In Figure 38 z-scans for fixed $y = 0$ are displayed. The uniform mode can be seen on the left part of this figure between 240 mT and 243 mT. Again, the origin of the great number of modes in this region is due to anisotropy [96].

Above 234 mT MSBVW have been detected. The modes propagate over a field range of 100 mT. For the analysis the first 50 modes have been

304 Spatially Resolved Detection of Microwave Absorption

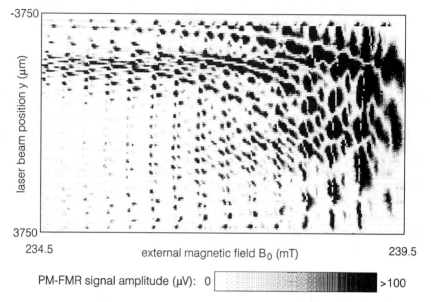

PM-FMR signal amplitude (µV): 0 [gradient] >100

Fig. 37 PM-FMR spectra of MSSW in the YIG film sample A. The heating laser beam was scanned along the y-direction. The microwave power = 0.4 mW and the modulation frequency = 1.1 kHz. The number of scans displayed here is 116.

Fig. 38 PM-FMR spectra of MSBVW in the YIG film sample A. The heating laser beam was scanned along the z-direction. The microwave power = 0.4 mW and the modulation frequency = 1.1 kHz. The number of scans displayed here is 67.

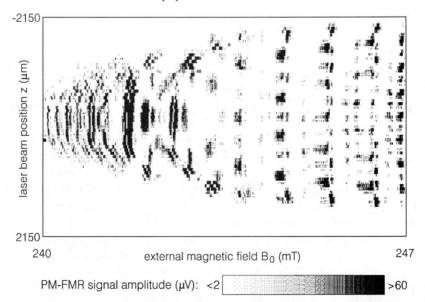

PM-FMR signal amplitude (µV): <2 [gradient] >60

used. The scan in the y-direction for the same values of B_0 in the case of the volume modes yields that n_y can be either 1 or 3.

In the spectra presented in Figs. 37 and 38 the intensities in the upper parts are more emphasized than those in the lower parts. This is due to the sample size and the position of the sample in the cavity. The sample cannot be placed symmetrically inside the cavity because at half height there has to be an annular ring. As the sample has to be placed in a way that it is entirely accessible by a laser beam, it has to be moved away from the center. The lateral dimension being 8 mm, a remarkable fraction of the cavity height, the rf field at the far end of the sample is already nonuniform which causes the described effect.

The measurements shown in this section have been obtained for a very low incident microwave power for which there is no distortion of the spectrum due to heating of the sample in the course of resonance absorption. The sensitivity of the photothermal experiments in this case is approximately one order of magnitude less than that of the conventional set-up [39].

With this photothermal detection scheme the temperature problem arising when passing through resonance as in the PD-FMR technique can be removed. Therefore, a new problem may arise: The local lateral resolution is strongly dependent on the focal size of the heating laser. With decreasing focus there is an increase in temperature rise in the focal spot on the sample which leads to a temperature rise of a few K for the whole sample.

For a rectangular geometry at theoretical solution of this thermal/electromagnetic problem seems to be impossible even under the assumption of numerous simplifications [32].

In order to study in more detail this problem the parameters that determine the PM-FMR signal of MSM in sample B, which allows perfect impedance matching of the resonant cavity, have been investigated.

One important result of these investigations is that the local temperature rise seems to be of no significance for the structure of the spectra. The individual lineshape offers the possibility to identify different contributions to the signal. With respect to the investigation of the lineshape temperature dependent measurements result in valuable new information. Nevertheless, there is, of course, a shift of the whole mode pattern due to the minor overall temperature rise within the sample, but neither in conventional spectra recorded with periodic laser perturbation, nor in the PM-FMR spectra could any evident differences be recognized.

With the smaller sample B PM-FMR spectra of MSM could be observed with a high SNR up to 120 kHz which is the actual frequency limit of the detection system. Based on this result an extention of up to 5 MHz should be possible [32]. The resolution of spatial patterns achieved with PM-FMR experiments with sample B was better than 50 μm at the moment. An even higher resolution down to 10 μm seems to be no problem.

4.5 Magnetostatic Modes in Spheres

The dispersion relation for magnetostatic modes in spheroids has been obtained by Walker [64]. A thorough discussion that explicitly states the resonance conditions and formulas for the rf field patterns for the special case of a sphere has been published by Fletcher and Bell [97]. Röschmann and Dötsch [98] have extended this work to higher-order modes and were the first to show detailed theoretical plots of the spatial distribution of rf magnetization patterns.

The outline given here is based on the Fletcher and Bell [97] and the Röschmann and Dötsch [98] papers.

The basic equations for the solution of the problem for spherical geometry are the same as for the case of a slab. Thus, inside the sample the magnetic potential ψ^i must satisfy

$$(1 + \kappa)\left(\frac{\partial^2 \psi^i}{\partial x^2} + \frac{\partial^2 \psi^i}{\partial y^2}\right) + \frac{\partial^2 \psi^i}{\partial z^2} = 0 \tag{78}$$

and outside Laplace's equation

$$\nabla^2 \psi^e = 0 \tag{79}$$

has to be fulfilled.

For circular polarization of the incident microwave the potential inside the sphere may be written as

$$\psi_n^m = P_n^m(\xi) P_n^m(\cos\eta) \exp(im\phi) \tag{80}$$

where ξ, η and ϕ are related to the cartesian coordinates by

$$x = a(-\kappa)^{1/2}(1 - \xi^2)^{1/2} \sin\eta \cos\phi \tag{81}$$

$$y = a(-k)^{1/2}(1 - \xi^2)^{1/2} \sin\eta \sin\phi \tag{82}$$

$$z = a\left(\frac{\kappa}{1 + \kappa}\right)^{1/2} \xi \cos\eta. \tag{83}$$

P_n^m is an associated Legendre function of the first kind [99]

$$P_\mu^m(\xi_0) = \frac{1}{2^n n!}(\xi_0^2 - 1)^{m/2}\frac{d^{m+n}(\xi_0^2 - 1)^n}{d\xi_0^{m+n}} \quad (84)$$

The solution of the boundary value problem is quite involved and leads to a dispersion relation given by [97]

$$n + 1 + \xi_0 \frac{P_n^{m\prime}(\xi_0)}{P_n^m(\xi_0)} \pm m\nu = 0, \quad (85)$$

where

$$\xi_0^2 = 1 + \frac{1}{\kappa},$$

$$\kappa = \frac{\Omega_H}{\Omega_H^2 - \Omega^2}, \quad \nu = \frac{\Omega}{\Omega_H^2 - \Omega^2},$$

$$\Omega_H = \frac{B_i}{M_s}, \quad \Omega = \frac{\omega}{\gamma M_s} \quad (86)$$

and

$$B_i = B_0 - \tfrac{1}{3}M_s$$

$$P_n^{m\prime} = \frac{dP_n^m(\xi_0)}{d\xi_0}.$$

The index n is a measure for the periodicity in the polar angle indicating sectors of reversal of the rf magnetization along the z-direction. m refers to the periodicity in the azimuthal coordinate thus implying $2m$ sectors in the z = const. plane where the rf magnetization reverses alternately.

It is convenient to classify the modes by the letter $s = n - m$. With this nomenclature a mode is then designated by $(m+s,m,r)$. Therefore, for fixed s subsequently the resonance equations are a function of m only. r indicates the number of the root of the solution. The labelling rises from $r = 0$ to $r_{max} = s/2$ for s even or to $r_{max} = (s - 1)/2$ for s odd. The modes with index $r = 0$ correspond to the roots at the highest fields. All modes lie within the frequency limits of $0 \leq \Omega - \Omega_H \leq 0.5$.

Modes with $s = 0$ and $s = 1$ show no dispersion. The uniform precessional mode is the (1,1,0) ($s = 0$) mode. This mode is degenerate with the (4,3,0) mode ($s = 1$) which is a remarkable result, especially from the point of view of nonlinear dynamics. In the coincidence region [100] a very efficient coupling between the two modes is assumed to be possible.

The excitation of nonuniform modes strongly depends on the distribution of the exciting rf magnetic field. The intensity of the excited mode as in the case of a slab is also dependent on the resulting transverse magnetic momentum so that modes with a greater number of variations in the magnetization are expected to have lower intensities [101].

One of the magnetization patterns will be explained in the course of the experiments performed on YIG spheres but for a detailed discussion of the behavior of the rf magnetization and the solution of the dispersion relation in spheres the reader is referred to Ref. [98].

4.6 Experimental Set-up for Spheres

The only photothermal method that is capable of obtaining spatially resolved FMR spectra of MSM on the surface and subsurface regions of spheres is the photothermally modulated technique. For this measurement the set-up described in sect. III.4.2 was used.

A restriction of this apparatus is the fact that the focal length of the lens used is rather short so that variations of the distance lens - sample of 1 mm (radius of the sphere investigated) already cause the focal spot diameter to increase from an optimum of 20 µm to 30 µm. In addition, the focal spot near the poles of the sample is no more a circle but is distorted elliptically due to the curvature of the sphere. From the experimental point of view it is no simple task to investigate a sphere in a cavity with the laser beam always impinging perpendicularly onto the sample surface. On the other hand, for the lower order modes with only some variations in the rf magnetization pattern in the sphere, the loss of resolution near the sample boundaries due to the non-perpendicular incidence of the laser beam is negligible.

This time, contrary to the PM-FMR experiments with films, the scans performed cover the whole surface of the sample and are not restricted to two perpendicular traces. As the resonance line position very sensitively depends on both the ambient temperature and the temperature in the sample which is effected by the resonance absorption itself, it is not possible to measure one value for a fixed external field. So at every point a spectrum over one resonance line had to be recorded. The typical linewidth is 0.1 mT and the scan has been performed over 0.4 mT. In an automated process the peak value of the PM-FMR signal and the corresponding phase angle have been extracted from the original spectra and are displayed as one point in the presented grey scale images. The step width between two laser positions was chosen as 50 µm which seems to be a good compromise between the lateral displacement of the sample and the thermal diffusion length at 1500 Hz (≈ 110 µm). So every plot of signal amplitude and phase consists of the results of 2025 individual scans, each lasting 30 s.

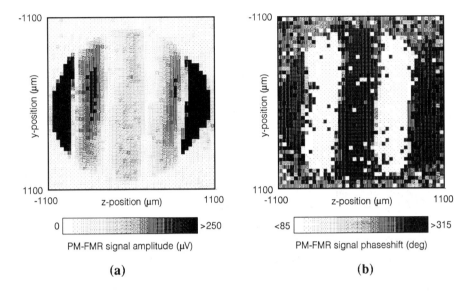

Fig. 39 PM-FMR signal amplitude (a,) and phase (b) of a YIG sphere of 2 mm diameter measured on a 50 μm grid.

4.7 PM-FMR Study of a 2 mm YIG Sphere

The sample was a polished YIG sphere of 2 mm diameter provided by our collaborators B. Benner and F. Rödelsperger from the Technische Hochschule Darmstadt. Conventional spectra of all excited modes (more than 30) have been recorded and several selected modes have been investigated by the PM-FMR technique [77]. In this chapter we are going to present only the results for one mode.

The identification of all recorded modes based on the theory mentioned above still faces some difficulties as the size of the sample under investigation is too large to ensure pure magnetostatic propagation. Fletcher et al. [101] found marked deviations of the measured line positions from the theoretical ones already for a 1.3 mm sphere. Moreover, these discrepancies depend on the mode index. In order to apply the procedure described in [101] to our special case the variable temperature equipment has to be extended for spatially resolved measurements, a configuration which is not available at present. The conventional investigation of the sample in a different cavity only does not yield the desired result as the distribution of the exciting rf magnetic field varies markedly and the spectrum does not include the same modes as the one measured with the light access cavity.

Nevertheless, the identification of the mode pattern shown in Fig. 39 is obvious. The observed resonance field coincides fairly well with that of a (3,1,0) mode calculated from the simple Walker theory. Our present understanding of the signal generation process supports this labelling. The theoretical rf magnetization pattern for a (3,1,0) mode is depicted in Fig. 40. The signal generation is determined by the superposition of two effects. The first one is the variation of the magnetization M_s of the sample due to the periodical heating. In the case of a uniform precession mode this produces the now well-known asymmetric lineshape of a PM-FMR signal. The contribution which arises from the distribution of the rf magnetization is superimposed on this signal. When the rf magnetization takes on its maximum value in the focal spot of the laser, the overall rf magnetization is decreased leading to a larger or smaller response of the sample to the incident radiation at the frequency of the impinging laser beam. The sign of the change depends on the temperature dependence of the line position which may either move to higher or lower fields when the temperature is raised depending on the mode under investigation.

When the phase is taken into account the observed strip pattern in Fig. 3.28 reproduces the one expected for a (3,1,0) mode. It consists of three maxima of which one is located in the middle and the other two at the margins of the sample. The extremely high amplitude of the PM-signal near the poles of the sphere is attributed to an agglomeration of magnetic dipole charges responsible for the homogeneous demagnetizing field. In that region the number of charges has to be higher than in the middle of the sample, so that an even bigger temperature derivative of the high-frequency susceptibility can be obtained.

Fig. 40 Distribution of the rf magnetization in the yz-plane for the (3,1,0) mode after [98].

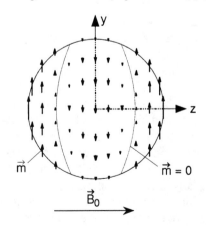

4.8 Conclusion

In the preceding sections the properties of magnetostatic modes in yttrium iron garnet (YIG) bulk and film rectangular samples and spheres were reviewed. It has been shown that the use of the PD- and the PM-FMR technique can provide useful information for the understanding of the dispersion relation, magnetization patterns and some new physical effects.

As far as the dispersion relation is concerned, the unique capabilities of the photothermal FMR detection schemes could be used to identify unambiguously, for example, anisotropy-induced volume modes, which from the resonance field position could have also been interpreted as surface modes. In this case the depth resolving capability was of great advantage. The spectra obtained with photothermal techniques could be analyzed quantitatively. The agreement for YIG films was even better than for bulk samples. This can be understood in the context of the electromagnetic problem, which at present has not been rigorously solved, yet it does not impose severe restrictions to thin samples.

The imaging of Walker modes in spheres, to the knowledge of the authors, is the first visualization reported ever.

Concerning new physical effects the high-power nonlinear behavior of the (1,5) volume mode in a YIG slab — which is attributed to a four-magnon decay — was discovered and shows a marked spatial variation. The agreement of physical parameters such as threshold power with nonlinear theory is good.

From the point of view of the photothermal measurement technique the results at high and low temperatures [32, 86] are very interesting as for both cases the signal amplitudes compared to the ones at ambient markedly grow. The temperature range covered by the PM-FMR experiment lies between 3.5 K and 600 K. The principal behavior of the temperature dependent data is well understood.

IV. SUMMARY

Three photothermal detection schemes in microwave absorption experiments and some of their promising applications to ferrimagnetic materials non-destructive characterization have been discussed. The common feature of all photothermal techniques is based on the utilization of the time and spatial dependence of the heat flow in the sample described by the concept of the thermal waves. Thermal waves are generated either by absorption of an intensity-modulated microwave radiation (photoacoustic effect and laser beam deflection) or by a supplementary intensity modulated laser beam incident on the sample exposed to the microwave field (phothermally modulated microwave absorption). The "time of flight" and the strong attenuation of the thermal waves deliver the tools to differentiate spatially the absorbing centers for the microwave radiation. The spatial resolution and the sensitivity depend on the particular detection scheme. Using the photoacoustic effect the average heat flow emerging from the sample is measured providing depth dependent but no lateral information on the heat sources. The depth resolution is limited by the time constant of the microphone which is typically a few hundreds of microseconds. During this time in the insulating ferrite materials the thermal wave propagates a distance of about 200 µm. The photothermal beam deflection does not suffer from such a limit in the time resolution of the detectors. Here the geometrical arrangement of the deflection set-up which has to be matched to the microwave cavity system imposes restrictions to the achievable signal intensity that predominantly determines the resolution limit of the laser beam deflection detection technique. Although the depth resolution in the case of the ferrimagnetic resonance measurements compares roughly with that of the photoacoustic detection, the laser beam deflection offers, in addition, a lateral resolution. This lateral resolving power exists in one dimension rather than in two dimensions as the signal is generated along the path of the laser beam on the sample surface. Sensitivity and resolution are again improved by the photothermally modulated microwave resonance absorption. Owing to the direct detection of the microwave and the possibility to focus the laser beam on the sample, the limits for the depth and the lateral resolution are improved by one order of magnitude. As compared to the laser beam deflection the photothermally modulated microwave technique provides a real three-dimensional tomographic method in the microwave spectral regime.

Despite the superiority of the photothermally modulated electronic detection, the other methods can provide advantages in special applications and for particular problems. The photoacoustic detection requires minor modifications of a conventional spectrometer and can be performed complementary to a conventional resonance measurement. The photoacoustic detection is the only one of the photothermal techniques that offers the possibility to calibrate the thermal response in order to determine absolute values of the absorption coefficients. Applied to ferrimagnetic materials the photoacoustically detected ferrimagnetic resonance has successfully been used to measure the inhomogeneous depth distribution of the microwave power density in soft ferrite slabs placed inside a waveguide. Lateral inhomogeneities of the microwave power density in the slab due to non-uniform demagnetizing fields could be resolved by means of the laser beam deflection. The demagnetizing field causes a magnetic field shift of the resonance line which varies with the lateral position. Ferrimagnetic resonance detected by laser beam deflection has also, for the first time, been used to visualize the spatial distribution of standing magnetostatic modes in a yttrium iron garnet slab. Also a spatially localized non-linear four-magnon process could be observed via the usual heat flow into one particular mode. Despite these promising examples there exists an important drawback of this technique. Owing to the fact that only a small number of the microwave absorbing centers contribute to the deflection signal, high microwave power densities are required which could modify the collective magnetic excitation due to locally inhomogeneous thermal loading of the sample. The most promising solution to this experimental problem is offered by the photothermal modulation of the microwave resonance absorption. Like in an experiment based on the conventional technique, the reflected microwave power is monitored, and therefore only small microwave power densities are required at the position of the sample. In addition, due to the high sensitivity of the conventional detection method rather low amplitude thermal waves have to excited by the modulated laser beam hitting the sample. Both conditions reduce the thermal loading of the sample considerably. Furthermore, the photothermally modulated detection scheme of microwave absorption provides the highest depth and lateral resolution, superior to those of the photoacoustic and of the laser beam deflection by at least one order of magnitude. This improvement is achieved by an increased complexity of the signal generation process. Whereas the PA- and the PD signals are directly proportional to the imaginary part of the high frequency susceptibility, the PM signal depends on the temperature derivative of the high frequency susceptibility. The superior imaging capability of the PM method has been demonstrated by photothermally modulated ferrimagnetic resonance experiments on YIG films and spheres. The results of these experiments also substantiate the need of spatially resolved resonance measurements. For the slab-shaped sample considerable deviations from the simple theory have been identified, for the first time, by the photothermal techniques in ferrimagnetic resonance.

V. REFERENCES

1. H.H. Perkampus, Naturwissenschaften, **69**, 162 (1982).

2. H. Vargas and L.C.M. Miranda, Physics Rep., **161**, 43 (1988).

3. A.C. Tam, Rev. Mod. Phys., **58**, 381 (1986).

4. S.E. Braslavsky, in *Photoacoustic and Photothermal Phenomena*, (P. Hess and J. Pelzl, Eds.) 508 (Springer Verlag, Berlin, 1988).

5. J.R. Barker and B.M. Toselli, in *Photothermal Investigations of Solids and Fluids*, (J.A. Sell, Ed.) (J. Wiley & Sons, Chichester, 1989).

6. J.C. Murphy, J.W.M. Spicer, L.C. Aamodt and B.S.H. Royce, Eds., *Photoacoustic and Photothermal Phenomena II.* Vol. **62** of *Springer Series in Optical Sciences.* (T. Tamir, Series Ed.) (Springer Verlag,

Berlin, 1990).

7. A. Mandelis, Ed., *Progress in Photothermal and Photoacoustic Science and Technology*. Vol. **1**. (Elsevier Science, New York, 1991).

8. M.J. Adams and G.F. Kirkbright, Analyst, **102**, 281 (1977).

9. A. Rosencwaig, *Physics Today*, (September issue, 1975) p. 23.

10. J.B. Aguayo, S.J. Blackbend, J.Schoenigr, M.A. Mattingly and M. Hintermann, Nature, **322**, 190 (1986).

11. S.S. Eaton and G.R. Eaton, in *Electron Spin Resonance*, **12B**, 176 (1991).

12. K. Ohno, Appl. Spectr. Rev., **22**, 1 (1986).

13. W. Wettling and W. Jantz, J. Magn. Magn. Mat., **45**, 364 (1984).

14. B. Neite and H. Dötsch, J. Appl. Phys., **62**, 648 (1987).

15. J.F. Dillon, Jr., L.R. Walker and J.P. Remeika, Proc. Int. Conf. Magnetism Nottingham, (1964).

16. S.M. Rezende and A. Azevedo, Phys. Rev., **B 44**, 7062 (1991).

17. C.J. Gorter, Physica, **3**, 503 (1936).

18. J.L. Dunn, C.A. Bates, M. Darcha, A. Vasson and A.M. Vasson, Phys. Rev., **B33**, 2029 (1986).

19. O.A. Cleves Nunes, A.M.M. Monteiro and K.S. Neto, Appl. Phys. Lett., **35**, 656 (1979).

20. C. Evora, R. Landers and H. Vargas, Appl. Phys. Lett., **36**, 864 (1980).

21. R.L. Melcher, Appl. Phys. Lett., **37**, 895 (1980).

22. U. Netzelmann and J. Pelzl, Appl. Phys. Lett., **44**, 854 (1984).

23. U. Netzelmann, U. Krebs and J. Pelzl, Appl. Phys. Lett., **44**, 1161 (1984).

24. T. Orth, U. Netzelmann and J. Pelzl, Appl. Phys. Lett., **53**, 1979 (1988).

25. J. Pelzl and U. Netzelmann, in *Photoacoustic, Photothermal and Photochemical Processes at Surfaces and in Thin Films*, (P. Hess, Ed.), Vol. **47** of *Topics in Current Physics* Ch. 12, 313 (Springer Verlag, Berlin, 1989).

26. J. Pelzl, U. Netzelmann, T. Orth and R. Kordecki, in *Photoacoustic and Photothermal Phenomena II*, (J.C. Murphy, J.W.M. Spicer, L.C. Aamodt and B.S.H. Royce, Eds.) 2 (Springer Verlag, Berlin, 1990).

27. D. Jiles, *Introduction to Magnetism and Magnetic Materials*. (Chapman and Hall, London, 1991).

28. T.L. Gilbert, Phys. Rev., **110**, 1243 (1955).

29. H.S. Carslaw and J.C. Jaeger, *Conduction of Heat in Solids*. (Clarendon Press, Oxford, 1990).

30. A. Rosencwaig and A. Gersho, J. Appl. Phys., **47**, 64 (1976).

31. A.C. Boccara, D. Fournier and J. Badoz, Apply. Phys. Lett., **36**, 130 (1980).

32. F. Schreiber, *Untersuchungen zur Signalentstehung und Ortsauflosung der photothermischmodulierten FMR in YIG*. Master's thesis, Ruhr-Universität, Bochum, 1992

33. U. Netzelmann, *Ortsaufgelöste Messungen der ferromagnetischen Resonanz mit Hilfe photothermischer Nachweismethoden*. PhD. thesis, Ruhr-Universität, Bochum, 1986.

34. T. Orth, *Photothermisch modulierte ferromagnetische und paramagnetische Resonanz*. Master's thesis, Ruhr-Universität, Bochum, 1988.

35. U. Netzelmann, E. v. Goldammer, J. Pelzl and H. Vargas, Appl. Optics, **21**, 32 (1982).

36. O. v. Geisau, U. Netzelmann and J. Pelzl, in *Photoacoustic and Photothermal Phenomena*, Vol. **58** of *Springer Series in Optical Sciences*, (P. Hess and J. Pelzl, Eds.) 451 (Springer Verlag, Berlin, 1988).

37. C.P. Poole, *Electron Spin Resonance*. (Interscience Publishers, New York, 1967).

38. R.F. Soohoo, *Theory and Application of Ferrites*. (Prentice-Hall, New Jersey, 1960).

39. L. Kreuzer, in *Optoacoustic spectroscopy and detection*, (Y.-H. Pao, Ed.) 1 (Academic, New York, 1977).

40. U. Netzelmann, J. Pelzl and D. Schmalbein, in *Fortschritte der Akustik*, Vol. **DAGA88**, 419 (1988).

41. U. Netzelmann, J. Pelzl and D. Schmalbein, in *Photoacoustic and Photothermal Phenomena*, (P. Hess and J. Pelzl, Eds.) 312 (Springer Verlag, Berlin, 1988).

42. J. Pelzl, Proc. 5th Conf. Acousto-electronics '91: (World Scientific, Singapore, 1992) 1.

43. W.B. Jackson, N.M. Amer, A.C. Boccara and D. Fournier, Appl. Opt., **20**, 1333 (1981).

44. M. Woelker, B. Bein, J. Pelzl and H. Walther, J. Appl. Phys., **70**, 603 (1991).

45. T. Orth, U. Netzelmann, R. Kordecki and J. Pelzl, J. Magn. Magn. Mat., **83**, 539 (1990).

46. M.A. Gilleo, in *Ferromagnetic Materials*, Vol. **2**, (E.P. Wohlfarth, Ed.) 2 (North-Holland, Amsterdam, 1980).

47. M. Hoffmann, O. von Geisau, S.A. Nikitov and J. Pelzl, J. Magn. Magn. Mat., **101**, 140 (1991).

48. U. Netzelmann, J. Pelzl, D. Fournier and A.C. Boccara, Can. J. Phys. **64**, 1307 (1986).

49. B. Lax and K.J. Button, *Microwave Ferrites and Ferrimagnetics*. (McGraw-Hill, New York, 1962).

50. R.S. Tebble and D.J. Craik, *Magnetic Materials*. (Wiley-Interscience, London, 1969).

51. D.L. Huber, in *Group III*, (K.-H. Hellwege and A.M. Hellwege, Eds.),

Vol. **4a** of *Landolt-Börnstein, New Series*, 315 (Springer-Verlag, Berlin, 1970).

52. H.P.J. Wijn, in *Group III*, (K.-H Hellwege and A.M. Hellwege, Eds.), Vol. **4b** of *Landolt-Börnstein, New Series*, 547 (Springer-Verlag, Berlin, 1970).

53. J. Smit and H.P.J. Wijn, *Ferrite*, (Philips Technische Bibliothek, Eindhoven, 1962).

54. P. Hansen, in *Physics of Magnetic Garnets*, 56 (North-Holland, Amsterdam, 1978). International School of Physics "Enrico Fermi", Course LXX (1977).

55. L. Thourel, *Depositifs a ferrites pour micro-ondes*, (Masson et Cie, Paris, 1969).

56. B.A. Auld, J. Appl. Phys., **31**, 1642 (1960).

57. M.H. Seavey and P.E. Tannenwald, *Electromagnetic propagation effects in ferromagnetic resonance*, Tech. Rep. 143, MIT Lincoln Laboratory (1957).

58. J.D. Jackson, *Classical Electrodynamics*, (New York: J. Wiley & Sons, 2nd Ed., 1975).

59. O. von Geisau, *Untersuchung der ferromagnetischen Resonanz in Ferriten mit photothermischen Nachweisverfahren*. Master's thesis, Ruhr-Universität, Bochum, 1987.

60. P.S. Epstein, Rev. Mod. Phys., **28**, 3 (1956).

61. C.L. Cesar, H. Vargas, U. Netzelmann and J. Pelzl, J. Magn. Magn. Mat., **54-57**, 1185 (1986).

62. R.I. Joseph and E. Schlömann, J. Appl. Phys., **36**, 1579 (1965).

63. R.L. White and I.H. Solt, Jr., Phys. Rev., **104**, 56 (1956).

64. L.R. Walker, Phys. Rev., **105**, 390 (1957).

65. R.W. Damon and J.R. Eshbach, J. Phys. Chem. Solids, **19**, 308 (1961).

66. L.K. Brundle and N.J. Freedman, Electron. Lett., **4**, 132 (1968).

67. G. Winkler, *Magnetic Garnets*, (Vieweg, Brauschweig, 1981).

68. A.G. Gurevich, *Magnetic Resonance in Ferrites and Antiferromagnetis* (in Russian) (Nauka, Moscow, 1973).

69. *Proc. IEEE*, Special section on "*Microwave Magnetics*", Vol. **76**, No. 2 (1988).

70. T.L. Carroll, L.M. Pecora and F.J. Rachford, J. Appl. Phys., **64**, 5396 (1988).

71. Y.T. Zhang, C.E. Patton and G. Srinivasan, J. Appl. Phys., **63**, 5433 (1988).

72. W. Wettling, W. Jantz and L. Engelhardt, Appl. Phys., **A 26**, 19 (1981).

73. N.P. Vlannes and F.R. Morgenthaler, J. Appl. Phys., **57**, 3721 (1985).

74. O. von Geisau, U. Netzelmann, S.M. Rezende and J. Pelzl, in MIOP '89 *Microwaves and Optronics Conference Proc.*, (K. Jansen, Ed.) (Hagenburg, 1989).

75. O. von Geisau, U. Netzelmann, S.M. Rezende and J. Pelzl, IEEE Trans. Magn., **26**, 1471 (1990).

76. F. Rödelsperger, O. von Geisau, H. Benner and J. Pelzl, Verhandl. DPG (VI), **27**, 682 (1992).

77. O. von Geisau, F. Rödelsperger, F. Schreiber, H. Benner and J. Pelzl, to be published, (1993).

78. A.D. Kusono and D.R. Tilley, J. Phys. C: Solid State Phys., **11**, 3487 (1978).

79. R. Ruppin, J. Appl. Phys., **62**, 11 (1987).

80. O.A. Chivileva, A.G. Gurevich and L.M. Émiryan, Sov. Phys. Solid State, **29**, 61 (1987).

81. A.V. Medved', I.P. Nikitin and L.M. Filimonova, Sov. J. Comm. Techn. Electr., **32**, 131 (1987).

82. I.V. Zavislyak, V.M. Talalaevskii and L.V. Chevnyuk, Sov. Phys.

Solid State, **31**, 906 (1989).

83. H. Suhl, J. Phys. Chem. Solids, **1**, 209 (1957).

84. A.D. Boardman and S.A. Nikitov, Phys. Rev., **B 38**, 11444 (1988).

85. V.E. Zakharov, V.S. L'vov and S.S. Starobinets, Sov. Phys. Solid State, **11**, 2368 (1970).

86. A.M.Mednikov, Sov. Phys. Solid State, **23**, 136 (1981).

87. A.V. Vashkovskii, V.I. Zubkov, E.G. Lokk and S.A. Nikitov, Sov. Phys. Solid State, **30**, 475 (1988).

88. P.E. Zil'berman, S.A. Nikitov and A.G. Temiryazev, JETP Lett., **42**, 110 (1985).

89. Y.V. Gulyaev, P.E. Zil'berman, S.A. Nikitov and A.G. Temiryazev, Sov. Phys. Solid State, **29**, 1031 (1987).

90. R.F. Belt, in *YIG Resonators and Filters,* (Wiley, Chichester, 1985) p. 226.

91. J. Helszajn, *YIG Resonators and Filters*, (Wiley, Chichester, 1985).

92. P. Görnert, Prog. Crystal Growth Charact., **20**, 263 (1990).

93. O. von Geisau, *Photothermische Untersuchungen der ferromagnetischen Resonanz kollektiver magnetischer Anregungen in YIG.* PhD. thesis, Ruhr-Universität, Bochum, (1993).

94. J. Barak, R. Ruppin and J.T. Suss, J. Appl. Phys., **63**, 2372 (1988).

95. R. Gieniusz and I. Smoczynski, J. Magn. Magn. Mat., **66**, 366 (1987).

96. M. Hoffmanu, *Untersuchung des räumlichen Verhaltens von magnetostatischen Moden in YIG-Filmen mit Hilfe der photothermisch modulierten ferromagnetischen Resonanz.* Master's thesis, Ruhr-Universiät, Bochum, (1993).

97. P.C. Fletcher and R.O. Bell, J. Appl. Phys., **30**, 687 (1959).

98. P. Röschmann and H. Dötsch, Phys. Stat. Sol. B., **82**, 11 (1977).

99. M. Abramowitz and I.A. Stegun, *Handbook of Mathematical Functions*. (Dover, 1965).

100. A.D. Boardman, Y.V. Gulyaev, S.A. Nikitov and W. Qi, in *Nonlinear Waves in Solid State Physics*, (A.D.Boardman *et al.*, Eds.) 315 (Plenum, New York, 1990).

101. P. Fletcher, I.H. Solt, Jr. and R. Bell, Phys. Rev. **114**, 739 (1959).

9

X-RAY THERMAL-WAVE NON-DESTRUCTIVE EVALUATION

Tsutomu Masujima[*] and Edward M. Eyring[+]

[*] *Institute of Pharmaceutical Sciences, Hiroshima University, School of Medicine, Kasumi 1-2-3, Hiroshima 734, Japan*

[+] *Department of Chemistry, University of Utah, Salt Lake City, UT 84112, U.S.A.*

I.	INTRODUCTION	322
II.	X-RAY ABSORPTION	322
III.	X-RAY PHOTOACOUSTIC SIGNALS AND INSTRUMENTATION	325
IV.	THE "SEMI-PULSE" X-RAY PHOTOACOUSTIC METHOD	328
	1. Depth Profiling	328
	2. Semi-Pulse Imaging Analysis	331
V.	X-RAY PHOTOACOUSTIC IMAGING	331
VI.	CONCLUSIONS	337
VII.	REFERENCES	337

I. INTRODUCTION

Photoacoustic spectroscopy has two special advantages for non-destructive evaluation (NDE) of materials. One is non-destructive spectroscopy which means that we can obtain the spectrum of a sample as is. The other advantage is non-destructive depth-profiling which means that we can get subsurface information by a non-destructive procedure.

Practically all photoacoustic studies of solids have been done in the UV-visible to IR wavelength region and not in the short wavelength region that includes X-rays. It is well known that microphonic photoacoustic techniques are effective at any wavelength of the electromagnetic spectrum that is absorbed by a sample solid and converted into heat [1]. The only major problem that one encounters in making photoacoustic measurements at X-ray wavelengths is in finding an X-ray source of sufficient intensity to make the photoacoustic signals emanating from the sample detectable. The severity of this technical problem is aggravated, if the X-ray source must also be wavelength tunable as in the extended X-ray absorption fine structure (EXAFS) method [2] for determining the composition of the first coordination sphere of a particular metal atom in a solid sample. The current solution to these problems is to secure time on an X-ray beam line of one of the higher energy synchotrons such as the Photon Factory (PF) at the National Laboratory for High Energy Physics in Tsukuba, Japan. Figure 1 provides a glimpse of this massive research facility. An interesting technical feature of this particular synchrotron is the substitution of positrons for electrons in the high vacuum ring from which the X-ray beams emerge tangentially. Positrons permit a longer duty cycle (about 12 hours) for the synchrotron between successive injections of charged particles at an acceleration energy of 2.5 GeV. The initial ring current after injections at the Photon Factory is approximately 350 mA.

II. X-RAY ABSORPTION

When materials are irradiated with X-rays several phenomena occur including elastic and inelastic X-ray scattering, fluorescent X-ray emission, generation of photoelectrons, and the appearance of Auger electrons. However, one further phenomenon must also exist: heat-generation. Photoacoustic detection is one of the sensitive methods of detecting weak heat generation. This fact suggested to one of us (T.M.) that an X-ray photoacoustic experiment would be feasible and should be attempted.

An X-ray absorption profile is shown in Fig. 2 which is unique in its abrupt increase of the absorption coefficient at the K-edge energy where the inner K-shell electrons are kicked out of their orbital. Since this inner energy value is uniquely dependent on the atom, identification of atomic species is readily accomplished. Furthermore, the comparative transparency of most

Fig. 1 Operation and control center (top) and view of beam lines BL14A to BL18 at the Photon Factory synchrotron in Tsukuba, Japan.

324 X-Ray Thermal-Wave Non-destructive Evaluation

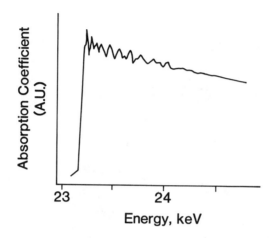

Fig. 2 Schematic of a (non-photoacoustic) X-ray absorption profile of an elemental solid showing the abrupt increase of the absorption coefficient at the K-edge energy.

Fig. 3 Photoacoustic signals of solid samples obtained with white X-rays. a) Cu (50 μm thick), b) Pb (400 μm), c) paper (100 μm), d) empty cell, i.e. no sample. Pulses at the bottom of each picture show the X-radiation (with intensity increasing in the downward direction).

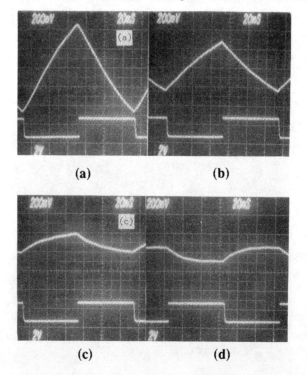

materials to a hard X-ray beam makes this method especially useful for NDE by the photoacoustic method.

III. X-RAY PHOTOACOUSTIC SIGNALS AND INSTRUMENTATION

Using synchrotron radiation as the X-ray source, the first preliminary X-ray photoacoustic experiments were carried out at Tsukuba. Figures 3 and 4 were the first signals from the X-ray photoacoustic phenomenon using white X-rays [3] and monochromatic X-rays [4], respectively. Even with monochromatic X-rays, weak heat generation and a subsequent temperature increase can be observed as triangular signal shapes. This temperature increase can be roughly estimated to be 10^{-4} to 10^{-5} °C. Since X-ray photoacoustic signals are very weak, the key device for experimental success is the photoacoustic sample cell. Figures 5 and 6 show a schematic interior view and a photograph of the exterior of a microphonic PA sample cell for detecting signals resulting from X-ray irradiation of samples. The sample cell interior chamber is cylindrical with a diameter of 10 mm and a height of 2 mm. A small electret microphone (Panasonic Omnidirectional Electret Condenser Microphone Cartridge P 9930) facilitates a compact sample cell design. Acoustic sealing of the back side of the microphone with a Teflon gasket is vitally important to obtaining a high signal-to-noise (S/N) ratio.

Fig. 4 Photoacoustic signals of solid samples obtained with monochromatic (λ = 1.56 Å) X-ray irradiation. Samples were the same as in Fig. 3.

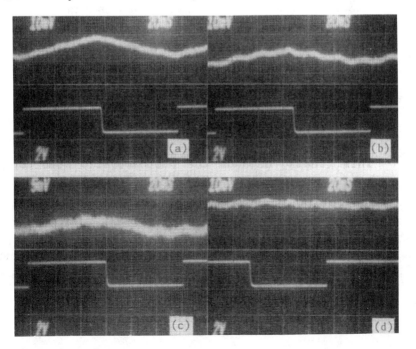

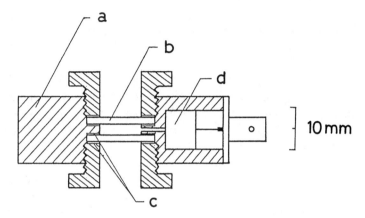

Fig. 5 Cross sectional schematic of an X-ray photoacoustic cell for solid samples. a) Cell body, b) beryllium window, c) Teflon gaskets, and d) microphone.

For gas phase samples a longer optical path (100 mm) and an isolated microphonic detector were adopted as shown in Fig. 7. The experimental set-ups for X-ray absorption spectroscopy and scanning X-ray photoacoustic imaging are depicted schematically in Figs. 8 and 9, respectively. Both systems are fully computerized with different intentions. For X-ray absorption spectroscopy [5], a microcomputer drives the monochromator which consists of two parallel-facing Si(111) crystals with sagittal focusing ability. This permits the operator to change the incident X-ray wavelength by changing the Bragg reflection angle of the monochromator. The microcomputer accumulates the two ion chamber currents thus giving the absorption coefficient and also the photoacoustic signal amplitude and phase.

In the case of scanning imaging [6], the X-ray beam is sharply focused on the sample at an approximately 1×1.5 mm^2 spot size but is not tunable in wavelength. Setting the aperture in front of the cell, spatial resolution can be adjusted. (0.6 ~ 0.8 mm spot diameters are common in our study). A microcomputer drives an X-Z stage for scanning the sample and also accumulates photoacoustic signal amplitudes and phase angles, both of which are normalized by the measured ionization chamber current to cancel the decrease in X-ray intensity that occurs in the course of the experiment. Photographs of the experimental set-up for photoacoustic EXAFS experiments at Beam Line 4A of the Photon Factory are shown in Figs. 10A and 10B. A typical example of a photoacoustic EXAFS X-ray absorption spectrum [7], that of $NiSO_4$, is shown in Fig. 11. The expected important future application of photoacoustic EXAFS will be in the non-destructive depth profiling of solid samples to depths of the order of tens of microns revealing the structure of subsurface layer materials. Since the photoacoustic signal becomes weaker

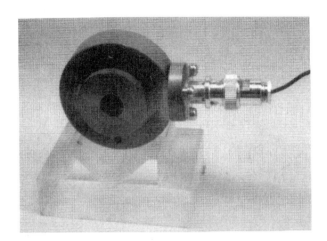

Fig. 6 X-ray microphonic photoacoustic sample cell.

with increasing depth of the subsurface sample structure, enhancement of the X-ray photoacoustic signal will have a high priority.

Ganguly and Somasundaram [8] demonstrated that volatile liquids such as diethyl ether placed inside a microphonic photoacoustic sample cell can enhance photoacoustic signals from solid samples by a factor of ten or more. The enhancement results from adsorption of the volatile liquid vapor on the

Fig. 7 Cross sectional schematic of an X-ray photoacoustic sample cell for gaseous samples. a,b) Window rings, c) beryllium windows, d) cell body, e) connectors, f) stainless steel tube and g) microphone.

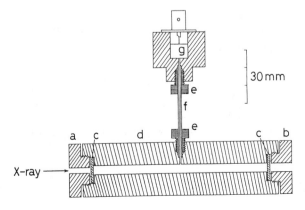

sample surface. A theoretical treatment of the enhancement including predicted photoacoustic signal amplitudes was published by Srinivasan, *et al.* [9]. The adsorbed liquid layer transfers heat between the solid sample and the ambient gas more efficiently. The signal is further enhanced by a pressure wave set up in the ambient gas by the periodic evaporation and condensation of the adsorbed molecules. The enhancement can be ten-fold or more at temperatures near, but below, the boiling point of the volatile liquid. Isak *et al.* [10] found that this enhancement was detectable but smaller when the solid sample was illuminated with synchrotron X-ray radiation rather than with visible light. They speculated that radiolysis of the volatile liquid by the X-rays was the cause of the decreased signal enhancement. The smaller enhancements were still sufficient to raise the X-ray radiation photoacoustic signal to a detectable level.

IV. THE "SEMI-PULSE" X-RAY PHOTOACOUSTIC METHOD

1. Depth Profiling

Synchrotron radiation is naturally pulsed at very high frequencies (on the order of a Gigahertz) that are unsuitable for microphonic photoacoustic measurements. The best S/N values for microphonic photoacoustic spectroscopy are achieved at audio frequencies of the order of 50 Hz as was shown by Rosencwaig and Gersho in their acoustic piston theory of photoacoustic spectroscopy [11]. Treating the high pulse repetition rate synchrotron radiation as if it were a continuous wave light source, one can chop the beam with a circular metal blade with two slits along its perimeter to create a 10 Hz to 50 Hz X-ray source. We tested this "semi-pulse" X-ray photoacoustic source by illuminating a layered sample. A 20 micron thick nickel metal foil was used as the sample base to which poly (ethylene terephalate) (PET) films of varying thicknesses were glued. The X-ray photoacoustic signal for a 75-micron thick PET film alone is shown in Fig. 12A and the photoacoustic signal from a layered sample of PET (75 microns thick) on top of a nickel foil (20 microns thick) is shown in Fig. 12B. Subtracting the PET signal shown in 12A from the layered sample signal shown in Fig. 12B one obtains the shape of the photoacoustic signal from the Ni foil sublayer shown in Fig. 13. A delay of 39 ms for the Ni signal is measured that can be used to give an estimate (97 µm) for the thickness of the PET layer (actually 75 µm). In principle, this type of measurement can be used to measure either the thickness of a layer in a sample or the thermal diffusivity of the layer material when one of the two pieces of information is already known.

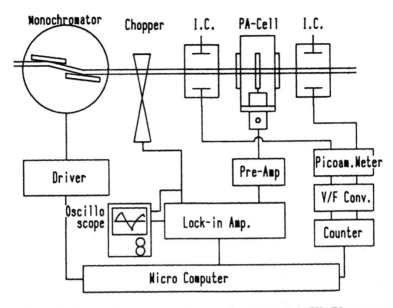

Fig. 8 Schematic diagram of an experimental set-up for photoacoustic EXAFS measurements.

Fig. 9 Schematic diagram of an X-ray photoacoustic imaging apparatus that uses a scanning focused X-ray beam.

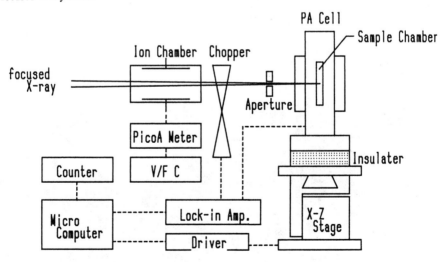

Fig. 10 Counter control system for the monochromator and photoacoustic EXAFS data gathering instrumentation outside the BL4A hutch (top). Detection unit (two ion chambers and a PA cell) inside the BL4A hutch (bottom).

2. Semi-Pulse Imaging Analysis

When we translate the X-ray beam point of irradiation linearly on the sample, we can obtain the semi-pulse response signal in sequence. The results obtained with a Ni foil sample that is partly covered by a polyethylene terephthalate (PET) film are shown in Fig. 14. Responses obtained at 14 points on the sample are shown in Fig. 14(a). The last five responses show a low level signal from the PET covered region. These signals are round in shape indicating the presence of more than one component. The last four responses were separated into two components as seen in Figs. 14(b) and (c). Figure 14 (c) shows that the secondary components were similar and indicates the existence of a homogeneous layer. The depth of the PET layer can be estimated by the delay time of the second component of the signal. Figure 14(d) shows the geometry of the PET film which generated these signals.

IV. X-RAY PHOTOACOUSTIC IMAGING

X-rays can be focused by a total reflection bending mirror. Using the focused beam from Beam Line 15A (at the Photon Factory) which was

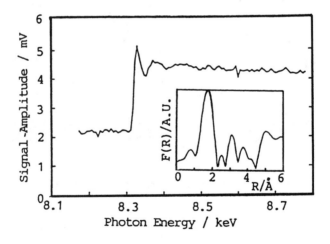

Fig. 11 Photoacoustic EXAFS of solid $NiSO_4$ with the radial distribution function in the insert.

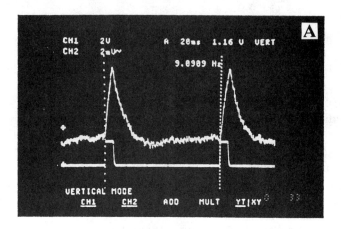

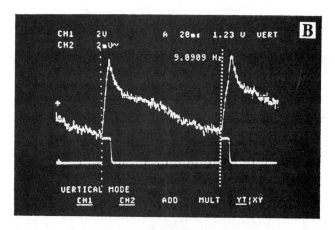

Fig. 12 A) X-ray photoacoustic signal from a 75 micron thick PET film. B) X-ray photoacoustic signal from a layered sample of PET (75 μm thick) glued on top of a nickel metal foil (20 μm thick).

originally designed for small angle X-ray scattering, imaging analysis has been attempted. Images at two X-ray energies were obtained, i.e. above and below the K-edge energy. Since the absorption coefficient of an atom changes drastically between these two energies, we can get a highlighted image of a specific element by subtraction of these two images. Figure 15 shows the images of photoacoustic signal amplitude obtained with a simple model sample. The absorption coefficient of Ni is quite different above (Fig. 15(b)) and below (Fig. 15(c)) the Ni K-edge but is essentially the same for Cu and Sn. Thus the subtraction image gives a highlighted region only in the Ni area. This method is quite useful for the analysis of dispersion of a

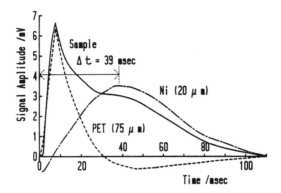

Fig. 13 Composition of the two signals contributing to Fig. 12(b) showing a 39 ms delay in the nickel foil signal.

specific elemental species. In this experiment the phase image was also obtained simultaneously. However, the result is similar in yielding the highlighted image of the Ni region. This similarity arises from the average depth of heat generation also being dependent on the absorption coefficient.

One more potential application of imaging analysis is three-dimensional NDE. This possibility was investigated with a model sample which was partly covered by PET. Calibration curves were obtained using simple model samples for signal amplitude vs. PET thickness and for phase angle vs. PET thickness. Signal amplitude measurements yield an exponentially decreasing curve with increasing PET thickness that goes down almost to zero at a PET depth of 300 microns. On the other hand, the phase curve shows a linear increase up to a 250 μm thickness of PET and shows saturation above this thickness. In both cases, a linear approximation can be applied until the PET thickness reaches 100 μm. After these calibration runs had been made, this method was applied to a more practical model sample as shown in Fig. 16(a). The signal amplitude image of Fig. 16(b) shows a high signal area in the Ni region, because the X-ray energy is just above the Ni K-edge. The phase image in Fig. 16(c) shows (in contrast to the amplitude image) a fairly level plot except for the PET coated region. This is because almost all of the sample is uncoated by PET and the signal is generated without a time-lag after the X-ray irradiation. The signal amplitude curve shows an exponential decrease and goes down almost to zero at a PET thickness of 300 microns. On the other hand, the phase curve increases linearly up to a PET thickness of 250 μm and then shows saturation above this thickness. In both cases a linear approximation can be applied until the PET thickness reaches 100 μm. The area 3, where 0.5 μm thick PET was glued by Epoxy onto the Ni foil, shows a phase delay at about 14°. The thickness of the PET layer can be estimated to be 10-16 μm for this delay according to the calibration curve, and is consistent with the actual value,

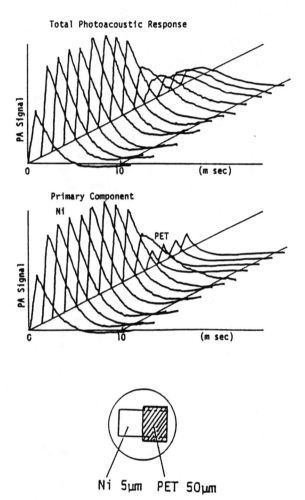
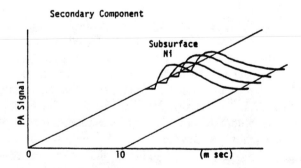

Fig. 14 Semi-pulse imaging analysis.

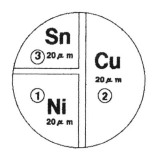

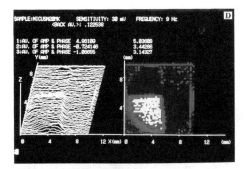

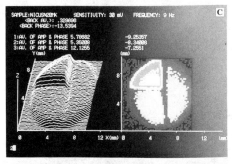

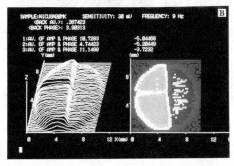

Fig. 15 Subtraction imaging of X-ray photoacoustic signal intensity. A) Configuration of a model sample, B) PA signal intensity image above the Ni K-edge, C) PA signal intensity below the Ni K-edge, and D) subtraction image, i.e. (b)-(c).

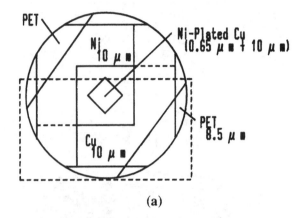

(a)

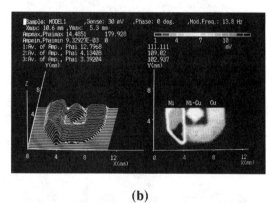

(b)

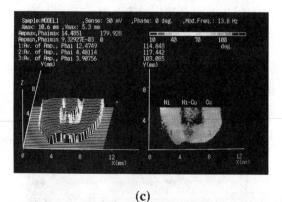

(c)

Fig. 16 a) Model sample for X-ray scanning imaging. b) X-ray scanning PA image of the sample shown in Fig. 16a based on PA signal amplitude. c) X-ray scanning PA image of the sample shown in Fig. 16a based on PA signal phase.

taking into consideration the glue layer.

The precise photoacoustic images of phase and amplitude seem to disclose the three dimensional distribution of materials. Self-consistent data analysis of phase and amplitude is most important for the evaluation of the depth by this non-destructive procedure.

VI. CONCLUSIONS

We have made efforts to apply the X-ray photoacoustic technique to practically interesting materials such as metalloenzymes in freeze dried powders. However, the sensitivity of the microphonic detection is presently insufficient to obtain clear spectra. To bring this method to a level where it will have practical or industrial applications will require a 10- to 100- fold improvement in sensitivity over what is now possible with the intense X-ray beam from a major synchrotron. The depth in a sample at which measurements can be made also depends on the X-ray beam intensity. It is certainly possible to increase the X-ray flux at a synchrotron using an undulator or multi-pole wiggler. However, a stronger X-ray beam causes greater damage to the sample.

Since materials are comparatively transparent to X-rays, the X-ray photoacoustic technique should eventually become an especially useful tool for depth profiling. For this reason major efforts to improve the sensitivity of the method and to develop the numerical analysis associated with depth profiling seem warranted. The dream driving these efforts is the achievement of three-dimensional, non-destructive evaluation of materials.

VII. REFERENCES

1. A. Rosencwaig, *Photoacoustics and Photoacoustic Spectroscopy*, (J. Wiley, New York, 1980).

2. B.K. Teo, *EXAFS: Basic Principles and Data Analysis*, (Springer-Verlag, Berlin, 1986).

3. T. Masujima, H. Kawata, Y. Amemiya, N. Kamiya, T. Katsura, T. Iwamoto, H. Yoshida, H. Imai and M. Ando, Chem. Lett., (Japan) 973, (1987).

4. T. Masujima, H. Kawata, M. Kataoka, M. Nomura, K. Kobayashi, M. Hoshi, C. Nagoshi, S. Vehara, T. Sano, H. Yoshida, S. Sakura, H. Imai and M. Ando, *Photon Factory Activity Report*, **5**, 139 (1987).

5. T. Masujima, H. Yoshida, H. Kawata, Y. Amemiya, T. Katsura, M. Ando, K. Fukui and M. Watanabe, Rev. Sci. Instrum., **60**, 2318

(1989).

6. T. Masujima, H. Kawata, M. Kataoka, H. Shiwaku, H. Yoshida, H. Imai, T. Toyoda, T. Sano, M. Nomura, A. Iida, K. Kobayashi and M. Ando, Rev. Sci. Instrum., **60**, 2522 (1989).

7. T. Masujima, H. Shiwaku, H. Yoshida, T. Sano, Y. Amemiya, H. Kawata, M. Kataoka, M. Hoshi, C. Nagoshi, S. Uehara, H. Imai and M. Ando, Rev. Sci. Instrum., **60**, 2468, (1989).

8. P. Ganguly and T. Somasundaram, Appl. Phys. Lett., **43**, 160 (1983).

9. J. Srinivasan, R. Kumar and K.S. Ghandi, Appl. Phys., **B 43**, 35 (1987).

10. S.J. Isak, B.A. Garland, E.M. Eyring, J.P. Kirkland and R.A. Neiser, Appl. Phys., **B 52**, 8 (1991).

11. A. Rosencwaig and A. Gersho, J. Appl. Phys., **47**, 64 (1976).

ACKNOWLEDGEMENTS

Chapter 5

The authors would like to acknowledge the General Electric Company for the loan of the infrared detector and for providing semiconductor samples. They would also like to thank Professor Sir Eric Ash for many helpful discussions.

Chapter 7

This chapter is dedicated to the late Dr. F.R. Thornley, a dear friend and valued colleague, who died of cancer in 1987. He was only 41. Richard contributed in no small measure to the development of the ideas and theoretical understanding behind *OTTER* and co-authored several early papers.

Chapter 8

The authors are indebted to F. Schreiber for critical reading of the manuscript and to F. Schreiber and M. Hoffman for providing unpublished data and stimulating discussions. They also thank H. Benner and F. Rödelsperger for helpful comments on the sphere measurements. Partial financial support by the Deutsche Forschungsgemeinschaft (Kennwort: Photothermik) is gratefully acknowledged. Parts of the work have been conducted within the frame of CAMST, Area B.

Chapter 9

Experiments were carried out at the Photon Factory, Tsukuba, Japan as Project Nos. 88-089, 90-076, and 92-095 and at the National Synchrotron Light Source, Upton, NY, U.S.A. as Project Nos. 91-X-279 and 89-X-26. Financial support from the Ministry of Education,Science and Culture, Japan [Nos. 02505001 and 01044098, International Scientific Research Program] and from the Department of Energy, Office of Basic Energy Sciences, U.S.A. is gratefully acknowledged.

Index

A

A-scan, 200
Absorption length, 199,200,214
Aircraft aluminum alloy panel, 40
Amorphous semiconductor superlattices, 152,165,167,175
Amorphous Si, 75,78ff,88,89
Annealing, 136

B

Black body, 112ff,136,140,144,149,193, 195
Black body radiation,112,114,120,125, 128,136,149
Bloch oscillator, 156
"Blue shift" phenomenon, 167
Body resonances, 267,272,274,282,283, 286,287
    conventional signal, 248,255, 259,283,287,295,302
    FMR measurements, 281,288
    heat conduction model, 246,275ff
    PA-FMR measurements, 283,284, 286
    PD-FMR measurements, 267,284, 286
    photothermal signal, 238ff 252,267,274,275,295
Box-car thermal wave imaging, 29,30,33, 34
Bulk temperature, 55

C

C-scan, 218
Cadmium mercury telluride, 112,128,146
Carrier diffusion length, 126,130,133,140
Carrier lifetime (determination of), 115ff 124,128,129,132ff
Carrier recombination, 119,133,139
Chemical vapor deposition, 139

Coating, 202ff,221,224,225
Composite, 214,224,226
Conduction band edge, 155,156
Contact holes, 103
Conventional FMR detection, 240ff
    sensitivity, 252ff,260
    setup, 247ff
    signal-to-noise ratio, 252
Crystalline superlattices, 152,157
Curie temperature, 258,262,263
Curing, 215
Cyclotron resonance, 172

D

Defects;
    detection, 4,5
    Si, 90
    Sub-surface, 29ff
Delaminations, 38
Demagnetizing field, 242,267,284,286,287, 296,310,312
Depth profiling, 200,322,326,328,337
Disbonds, 38
Dislocations;
    Si, 91,95
Dispersion relation for thermal waves, 25
Displaced atom density, 78
Drying, 215,216

E

Effective magnetic field, B_{eff}, 242
Eigenvalues, 157
Electrical resistivity, 157
Electro-modulation reflectance, 167
Electron paramagnetic resonance (EPR), 240,244ff,279
Electron-hole pairs, 113
Electronic transport and optical properties, 152
Emission spectroscopy, 209
Emissivity (measurement of), 114,142

Emissivity, 112,120,124,128,136,140ff 193,195,204,216
Energy band structure, 156,157
EPR spectrometer, 279,294
Equation of motion of magnetization, 289
Etch monitoring, 85
Excitation spectroscopy, 207
Excitonic structures, 159
Extended X-ray absorption fine structure (EXAFS), 322

F

Ferrites, 261ff
    crystal structure, 262
    electromagnetic propagation, 261,265,267,268
    exchange propagation (spin waves), 266
    high-frequency properties, 241,261,263
    magnetic properties, 240,256,257,262,263
    magnetostatic propagation, 266,309
    plane wave propagation, 263ff
Ferromagnetic resonance (FMR), 240-303
Film, 6ff,10,203ff,211ff,221,222,226
Flying spot thermal wave camera, 42,44,46
Fourier transform, 56,59,125,126

G

GaAlAs, 138ff
GaAs, 115,124,129,136,138ff
Gallium arsenide, 138
Gas microphone, 4,5,9
Generation of photoexcited carrier waves, 116
Generation of thermal waves, 119
Grain boundaries, 88
Graphite-epoxy laminates, 35
Grey body, 195

H

Heat diffusion equation, 246
Heat spread function, 38,39
Heteroepitaxy, 152
Hexaferrites, 262
High dose implants, 82
Homojunctions, 156
Hot-wall epitaxy (HWE), 152
Hydrogenated amorphous semiconductors, 152

I

Impulse excitation, 189,191,195,196,203, 205,209,213,218,221
Impulse response, 189,195ff,200,205,206, 220
Index of refraction, 161,179,180
Infrared absorption, 121ff
Infrared detector, 122,123,125,128ff
Infrared emission, 123,124,141,142,146, 147
Inhomogeneous demagnetizing field, 267,284,286,295,296
Inverse scattering and object reconstruction, 38
Ion implant monitoring, 77
Ion implantation, 115,133,134

K

Kirchoff's law, 120-122,193

L

Laser diode, 128
Lattice damage, 77,78,85,86
Lattice disorder, 135,136
Layered sample, 54,64
Layers, 6,7,10,64ff
Liquid phase epitaxy (LPE), 152,159
Lock-in video thermal wave imaging, 40
Low-pass filtering, 39

M

Magnetic excitations, 239,240,263
Magnetic permeability, 241
Magnetic resonance line width, 241,244
Magnetic susceptibility, 241
Magnetostatic modes (MSM), 261,262, 265,288ff,306,311,312
    backward volume wave modes (MS-BVW), 291
        experiment, 296
        PD imaging, 297
        PM imaging, 303ff
    dispersion relation, 261,265,266,291,296,298,306ff
    experiment, 303
    magnetocrystalline anisotropy, 257,293
    nonlinear effects, 293
    retarded modes, 293
    surface wave modes (MSSW), 292,293.296ff
        experiment, 297
        PD imaging, 297
        PM imaging, 303ff
    Walker modes, 288,311
        dispersion relation, 261,265,266,291,296, 298,306ff,311
        PM imaging, 309
Metal defects, 91
Metal precipitates, 91,95
Metal thickness, 86
Metal-organic vapor phase epitaxy (MOVPE), 152
Metallization monitoring, 86
Microphonic detection, 337
Microwaves, 8,9
Modulated reflectance, 6,8
Modulated reflectance apparatus, 75
Moisture, 214-217
Molecular beam epitaxy (MBE), 152

N

Nonradiative recombination, 166

O

One-dimension, 54,56,57,62,65,67ff
Optical absorption spectra, 165ff
Optical band gap, 157
Optical beam deflection, 5,8
Optical limit, 196,200

P

p-n junction, 121,136,159
Particle beams, 8,9
Phase transition, 190,218
Phonon softening, 167
Photoacoustic EXAFS, 331
Photoacoustic FMR detection (PA-FMR), 246,250,253,260,283
Photoacoustic imaging, 326,329,331
Photoexcited carrier plasma, 118
Photogenerated carriers, 166,167
Photoluminescence excitation (PE) spectroscopy, 153
Photothermal deflection FMR detection, 246,254ff,261,284
Photothermal displacement, 135
Photothermal radiometry (continuous modulated), 112ff
    for semiconductor analysis, 114
    microscope, 126ff
    of induced carrier properties, 113,116ff
    of thermal properties, 112
    signal generation, 116ff,119
Photothermal radiometry, 5
Photothermally modulated FMR detection (PM-FMR), 255,258ff,274ff
Physical properties, 7,8,10
Picosecond laser ultrasonics, 179
Piezoelectric, 3ff,9
Piezoelectric transducer, 157,159
Planck's function, 193
Planck's law, 120,122
Plasma chemical vapor deposition (PCVD), 152,153,161,165,167
Plasma wave interference, 80
Polycrystalline Si, 75,88
Potential barrier, 155
Potential wells, 152,155ff,166,167
Propagation coefficient, 118,125

Q

Quantum confinement effect, 159
Quantum effect/quanization of energy levels, 156,157
Quantum efficiency, 156
Quantum mechanics, 156
Quantum size effect, 152,157,163,166,167
Quantum wells/quantum well wire, 153,156,159,160,168,171,172

R

Radiative recombination, 167
Radio frequency waves, 8,10
Reactive ion etching, 115,138
Reflecting objective, 127,128
Resonance tunelling effect, 156
Resonant raman scattering, 153
Rutherford backscattering, 77,85

S

Scanning, 189,218,220ff
Scattering, 202,210,211,214
Semi-pulse X-ray photoacoustics, 328
Semiconductor assessment, 129
Semiconductor superlattices and heterojunctions, 152,153,180
    Photo-modulation reflectance (PE), 167ff
    Photothermal deflection mirage effect, 160
    Spectrophotometer, 165
    Photoacoustic (PA) and photothermal (PT) techniques, 157ff
Silicon membrane, 144,145
Sinusoidal, 189,192,195,200,221
Soft ferrites, 260
Solution of the heat equation with a pulsed source, 25
Solution of the heat equation with a moving source, 27
Solution of the heat equation with a periodic source, 25,55ff
Spatial fast fourier transform (FFT), 38

Spatial resolution, 39,115,126,147
Spin waves, 293
Stefan's law, 193
Stefan-Boltzmann law, 112,120
Step, 190,192,195,199,204,215,216,220
Strained heterolayer superlattices, 152
Subgap absorption spectra, 160ff
Surface acoustic wave, 180
Surface recombination (effect of), 118,130,138
Surface recombination, 114,117,120,130,132,138,146
Surface recombination velocity (determination of), 133,140
Synchrotron radiation, 325,328

T

Temperature field, 186ff,195,201
Thermal conductivity, 174ff
Thermal diffusion length, 54,56,192,199, 200,205,206,211,216,246,253,254,256, 257,260,284,296
Thermal diffusivity, 167,173ff,177,180, 187,189,201,203,205ff,211,213
Thermal effusivity, 187,198,216
Thermal limit, 196,198,200,201
Thermal propagation wavenumber, 55
Thermal resistance, 213,214,218,221
Thermal resistivity, 175
Thermal wave, 186,187,189,191,201,202, 206,209,213ff
Thermal wave echo, 29,30,32,33,35ff
Thermal wave imaging, 24,29ff,38,40, 91ff,95,98,100
Thermal wave tomography, 35
Thermoacoustics, 3,4
Thermoelastic deflection, 62
Thermoelastic displacement, 59,65,67
Thermography, 186,224
Three-dimensions, 54,56,57,58,62ff
Total displaced volume, 61,66,68,69
Tunelling devices, 156
Turbidity, 210,211
Two dimensional electronic gas (2DEG), 171

U

Ultrasonic, 36
Ultrasonics, 3
Urbach edge/exponential absorption edge, 160

V

Valence band edge, 155,156
Voids, 90,99ff,106

X

X-ray photoacoustics, 322,325ff,331ff
X-Rays, 8,9

Y

YIG, 256ff,262,288ff,311,313
Yttrium iron garnet (YIG), 256,257,259,288
    crystal structure, 262
    samples, 293,295,298, 300,302,304